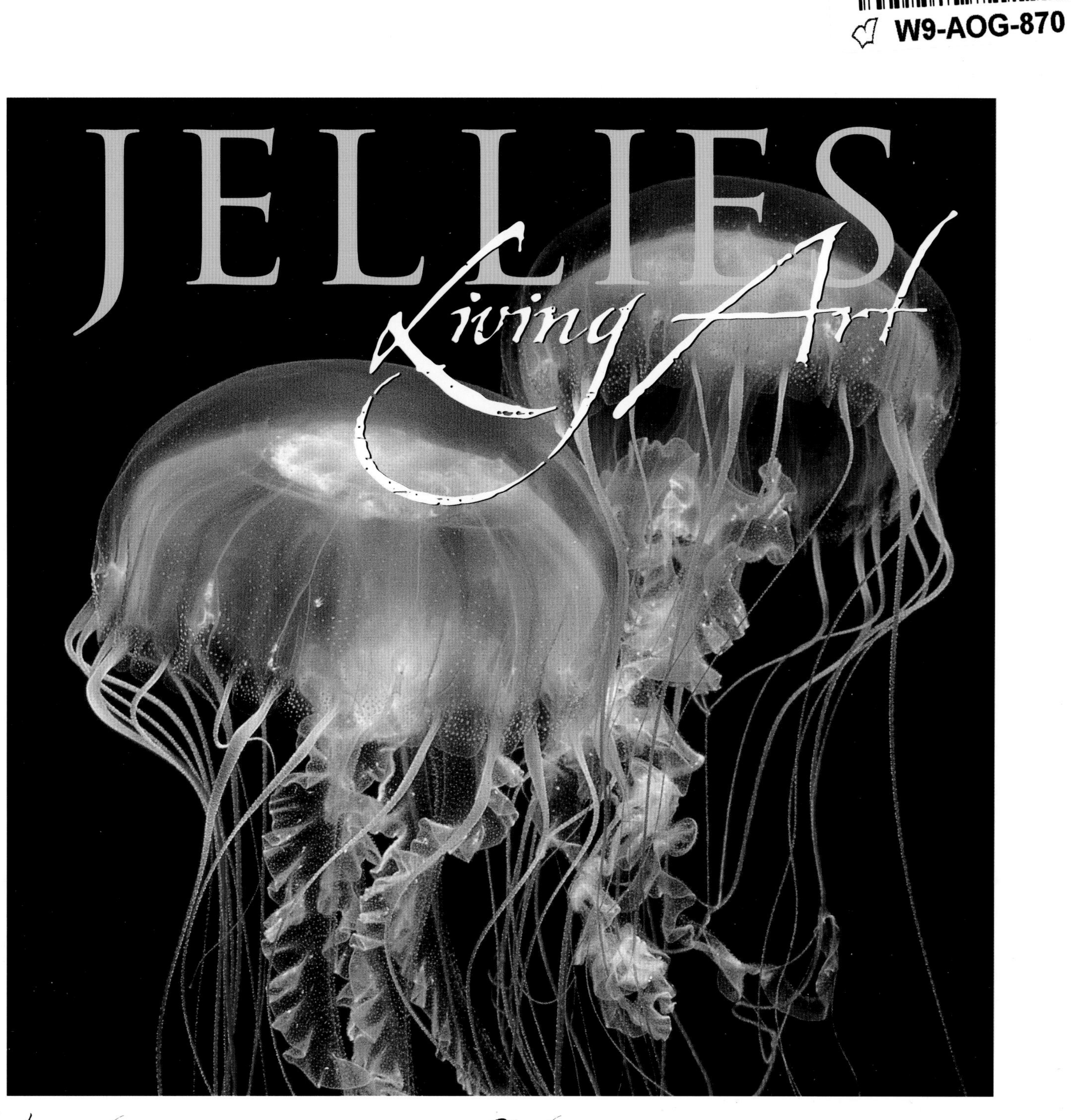

JELLIES: Living Art

Foreword by TERRY TEMPEST WILLIAMS Written by JUDITH L. CONNOR & NORA L. DEANS

MONTEREY BAY AQUARIUM

To Peter, my beloved northern star.
—NLD

To Michael and Alexander, my moon, my son, la luz de cada día.
—JLC

ACKNOWLEDGMENTS

This book came to be in part thanks to the inspiration of artist, writer, exhibit developer and friend Jaci Tomulonis, whose vision gave rise to the innovative exhibit, Jellies: Living Art.

With deep gratitude to Terry Tempest Williams for encouragement and inspiration; to our publisher, Michelle McKenzie and our dear editor, Miki Elizondo for their careful, thoughtful way with words; to Diane Tempest and the delight of first discovery.

Published in the United States by the Monterey Bay Aquarium Foundation, 886 Cannery Row, Monterey, California 93940-1085
www.montereybayaquarium.org

ISBN: 1-878244-38-8

Library of Congress Cataloging-in-Publication Data:
Connor, Judith, 1948—
Jellies : living art / written by
Judith L. Connor, Nora L. Deans.
p. cm.
ISBN 1-878244-38-8
I. Jellyfishes. I. Deans, Nora L. II. Title.
QL377.S4 C73 2002
593.5'3--dc21
2001007751

MANAGING EDITOR: Michelle McKenzie
PROJECT EDITOR: Miki Elizondo
EDITOR: Lisa Tooker
DESIGNER: Elizabeth Watson
CONTRIBUTING WRITERS: Jaci Tomulonis, Elizabeth Labor, Melissa Hutchinson/art and artist captions

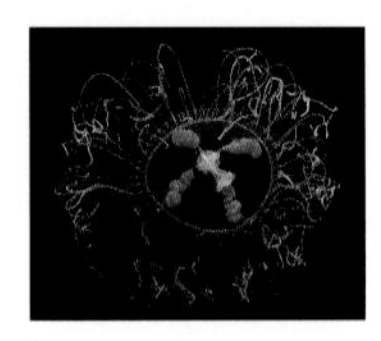

PHOTO CAPTIONS: Front cover and page 1, Sea nettles, *Chrysaora fuscescens*; Above, The jewel-like hydromedusa, *Vallentinia adherens*, lives hidden in seaweeds; Page 3, Eight curved gonads dangle below the transparent bell of *Crossota alba;* Pages 4-5, The tentacles of these sea nettles, *Chrysaora fuscescens*, look like lacy ribbons, but pack a potent sting; Pages 6-7, The black sea nettle, *Chrysaora achlyos*, displays the classic medusa shape, with rounded bell, long tentacles and ruffled mouth-arms.

PHOTO CREDITS: Front Cover: Craig W. Racicot/MBA; Back Cover: David Wrobel/MBA; Front Flap: George Matsumoto/MBARI; Authors' Photo: Michelle McKenzie; Bob Cranston/seapics.com: 29; Ben Cropp: 49; Bill Curtsinger: 17; Steven Haddock: 3, 15, 18, 23, 48, 70; Howard Hall/howardhall.com: 58-59; Tim Hellier/ imagequest 3d.com: 52; Richard Herrmann: 19; Jay Ireland & Georgienne; Bradley/bradleyireland.com: 25 (top left); David Kearnes: 28 (lower left); Larry Madin: 89; George Matsumoto/MBARI: 14, 46, 67, 77, 88; Claudia Mills/MBA: 32; Dawn Murray/MBARI: 40; Chris Parks/imagequest 3d.com: 26; Peter Parks/ imagequest 3d.com: 16, 25 (right), 84; Kevin Raskoff/MBARI: 8, 12, 42, 50, 93; Elenora de Sabata/ seapics.com: 9; Becca Saunders/Auscape: 61 (top left); Rob Sherlock/MBARI: 44,45; Tom & Therisa Stack/Auscape: 64; Scott Tuason/ imagequest 3d.com: 82-83; Randy Wilder/MBA: 4-5, 34-35; David Wrobel: 71; David Wrobel/MBA: 2, 6-7, 10, 11, 20, 22, 24, 28 (top), 30, 31, 36, 37, 38,41,43, 47, 51, 53, 54, 57, 61 (right), 63, 65, 66, 68-69, 72, 73, 74 (top and bottom), 75, 78-79, 80, 86 (left and right), 90, 95; Marsh Youngbluth: 94.

LITERARY CREDITS:
Rachel Carson, *The Edge of the Sea.* Text © 1955 by Rachel L. Carson, renewed 1983 by Roger Christie. Reprinted by permission of Houghton Mifflin Company.
Rachel Carson, *Under The Sea-Wind.* New York: Penguin Books USA Inc., 1996. © 1941 by Rachel Carson, renewed 1969 by Roger Christie. Illustrations copyright Bob Hines, 1991.
Robinson Jeffers, "Fog", *The Selected Poetry of Robinson Jeffers.* Edited by Tim Hunt. © 1927, 1928, 1938 by Robinson Jeffers, renewed. With permission of Stanford University Press.
Raquel Jodorowsky, " Malaguas", *Woman Who Has Sprouted Wings: Poems by Contemporary Latin American Women Poets.* Translated by Pamela Cornell, edited by Mary Crow. Pittsburgh, PA: Latin American Literary Review Press, 1984.
Pablo Neruda, "Enigmas", *Pablo Neruda, Five Decades: A Selection (Poems 1925–1970).* Edited/translated by Ben Belitt. New York: Grove Press, 1974.
Pablo Neruda, "Oda a las Algas del Océan", *Selected Odes of Pablo Neruda.* Edited/translated by Margaret Sayers Peden. © 1990 Regents of the University of California, © Fundacion Pablo Neruda.
Edward F. Ricketts, Jack Calvin and Joel W. Hedgpeth, *Between Pacific Tides.* Foreword by John Steinbeck. © 1939 by the Board of Trustees of the Leland Stanford Jr. University; renewed 1948, 1952, 1962, 1968, 1985.
John Steinbeck, *The Log from the* Sea of Cortez. © 1941 by John Steinbeck and Edward F. Ricketts. © renewed 1969 by John Steinbeck and Edward F. Ricketts, Jr. Used by permission of Viking Penguin, a division of Penguin Putnam, Inc.
Terry Tempest Williams, *Leap.* New York: Pantheon, Random House, Inc. 2000.

Printed on recycled paper and bound in Hong Kong by Global Interprint.

C O N T E N T S

PARTNERSHIP

No conocéis tal vez	*You may not know*
las desgranadas	*the spilling beads*
vertientes	*of the slopes*
del océano.	*of the ocean.*
En mi patria	*In my homeland,*
es la luz	*ocean is the light*
de cada día.	*of each new day.*
Vivimos	*We live*
en el filo	*on the edge*
de la ola,	*of the wave,*
en el olor del mar,	*with the smell of the sea,*
en su estrellado vino.	*with its starry wine.*

—Pablo Neruda, from *Oda a las Algas del Océano*

Twenty years ago, we were siblings who hadn't met. Work and nature brought us together. The work was sometimes a struggle for each of us—a struggle for time and words, a struggle to understand and describe the natural world. Of all the themes we tackled, jellies proved to be the topic of transparency and contrast, clear and

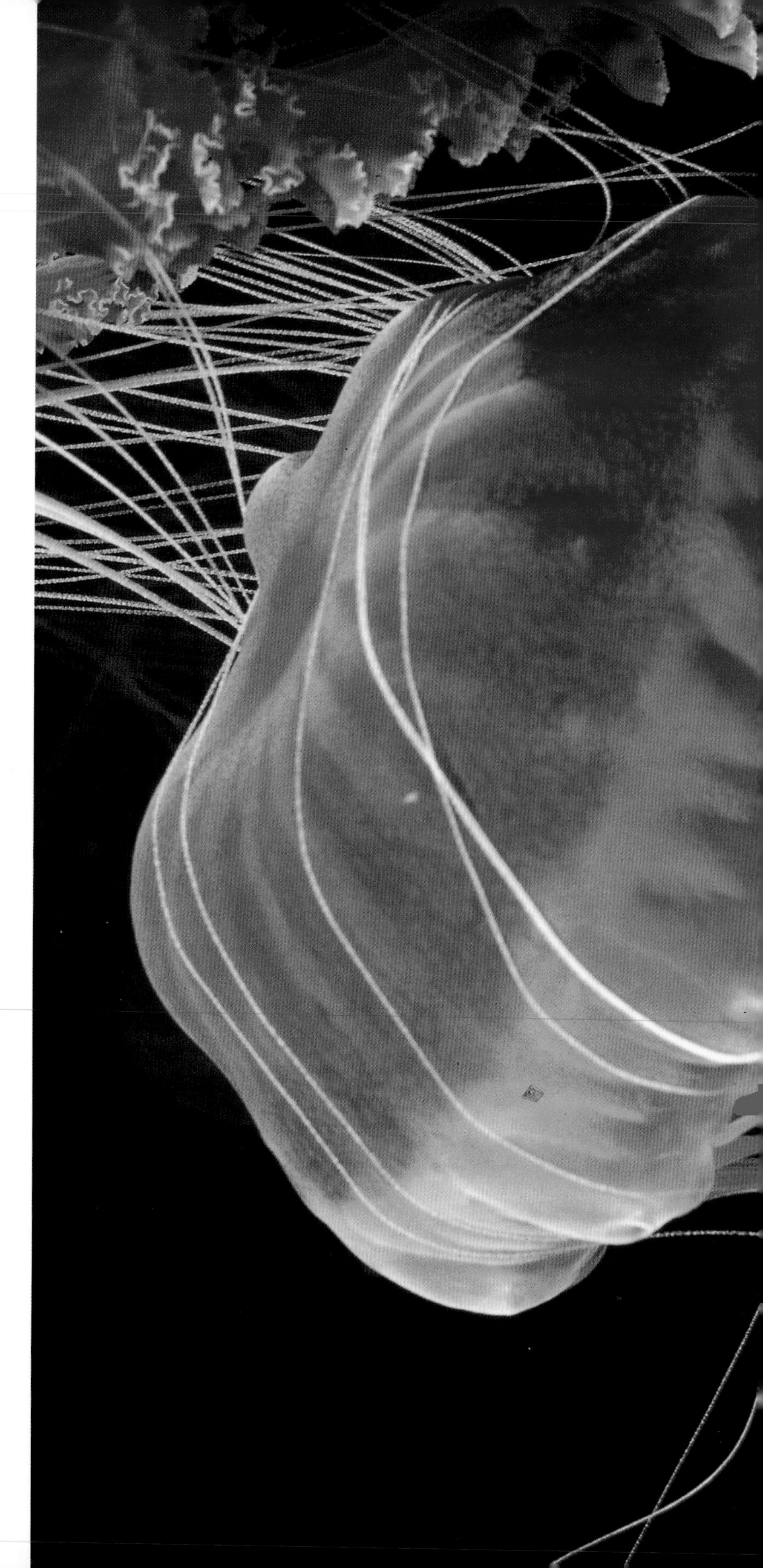

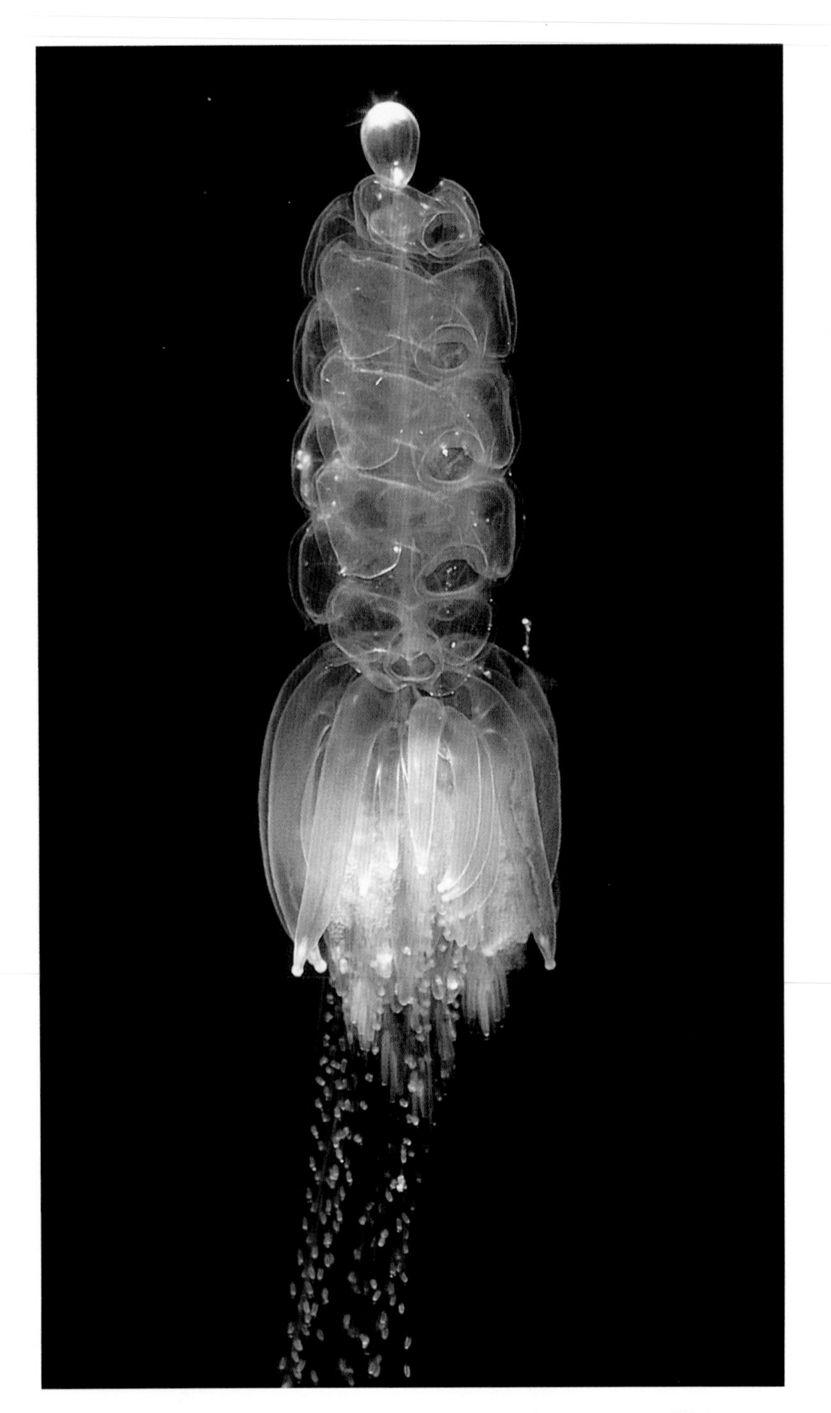

The siphonophore, Physophora hydrostatica, *bears a tiny gas-filled float above its cluster of swimming bells.*

free-floating, their guts and gonads on display for anyone to see. Jellies—sometimes floating with the current, sometimes pumping their way upstream—became a metaphor for our work: writing, conservation, science and art.

Close friends brought together by a love of words and wildlife, we weave together our different perspectives, an alliance of scientist and artist. Our styles diverge even to how we identify each other—Nora by first name; Connor by surname. In the early days of our friendship, our writing tables were wedged together in a tiny Monterey office. Now thousands of miles separate one's desk from the other's lab table.

The sea and languages are our common bonds. For years, we've shared favorite books, photographs, ideas and writings, seeking inspiration and honest criticism. Our long-distance friendship defies the odds; it persists in spite of time, distance and our disparate lifestyles.

Over those years, we witnessed the birth of sister institutions: an aquarium and a research institute. We were lucky to be part of the early days of the Monterey Bay Aquarium and the Monterey Bay Aquarium Research Institute. Like the institutions themselves, we each had our distinct focus and approach, but shared the goal of learning more about nature and the oceans. In our work, we planned exhibits, taught classes and wrote books. Sometimes in familial collaboration, sometimes in solitary terror, we stretched our minds around new topics: film, fisheries, art, law, seaweeds, indigenous cultures and software code.

We watched as our institutions matured under the leadership of two amazing women, Julie Packard and Marcia McNutt. They took risks, launched new programs and guided us toward a vision of ocean exploration and conservation. We saw the first kelp forest behind glass. We used robotic vehicles to probe the ocean depths. We glimpsed a satellite's view of our blue planet. We studied worms and whales, seaweeds and snails. But it was the jellies, the gelatinous ocean animals, that showed us living art in motion.

SUN AND MOONS IN MOTION

by TERRY TEMPEST WILLIAMS

Of the sea. Open sea. Transparent. Translucent. Transcendent. Bell and tentacles. Nothing hidden. A gelatinous body of nerves. Pulsating. Throbbing. Drifting. A jelly is more verb than noun.

What is it? Why is it that when we enter the shrine of jellies in the Monterey Bay Aquarium, we enter a state of awe and reverie? Our voices become hushed. Our eyes soften. We step closer to each exhibit as if to witness for ourselves the mind of the Holy. We are touched. The pulse of the jellies becomes our own as we feel a rhythmic beauty enter our own veins. Our blood pressure lowers. We make vows to slowness and feel the weight and encumbrance of flesh and bones.

I recall a day in Kachemak Bay, Alaska. My husband and I were kayaking the emerald sea. We were aware of otters floating on their backs breaking open urchins with stones, seals rising and falling with the currents, and bald eagles perched on snags, breaking the silence with their high-pitched chirps. The steep slopes of the rainforest on either side of us held us in place within this narrow inlet. On this particular afternoon, the water was like glass. We dipped our paddles into the sea gently, hesitant to break the water's surface. As I leaned over ever so slightly, I suddenly noticed that what was below us was in stark contrast to what was above. A fast-moving current of moon jellies, hundreds, maybe even thousands of transparent bowls of light were streaming by. My first impulse was to put my hand in the living water to touch them, hold them, a moon in my hand. But I held back, my childhood was full of stinging moments on California beaches where the long reach of tentacles burned our skin in search of prey.

The moon jelly, Aurelia aurita, *appears in Monterey Bay in autumn.*

And so Brooke and I watched, for hours we watched, a bloom of *Aurelia*, moon jellies, unfurl in the frigid waters until twilight forced us back to land.

Is it enough to know that true jellies, or medusae, belong to the Phylum Cnidaria, along with corals and anemones, that they can sting and stun their prey with tiny structures called nematocysts making them among the most efficient and deadly predators of the sea?

Or that the consciousness of a jelly is not privy to what state of development has preceded its present life or what will follow, be it egg or polyp or a free-swimming ephyra with eight tiny arms radiating outward like a star?

Or what about the siphonophores, linear colonies of individuals connected along a common stem, that sometimes reach 30 meters long, appearing at times like an endless strand of Christmas lights drifting in deep waters?

Or consider the comb jellies, jewel-like in appearance, belonging to the Phylum Ctenophora, who manage locomotion through vibrating cilia, preying on zooplankton. They can become almost invisible against the backdrop of "marine snow," the blizzard of detritus that falls through the sea.

We can wonder how the success of jellies within the framework of both form and formlessness has made them among the most successful of organisms, their fossil record on this planet recording some 700 million years.

But it is their overriding aesthetic that brings me to my knees as it has done on so many occasions. One morning while beachcombing, also in Southeast Alaska, I came across a perfectly drawn white cross, centered inside a circle, also white. It appeared on the mudflat like a medicine wheel. Only later, did I discover what I had seen—a signature left by the proteinaceous radial canals of *Mitrocoma cellularia*. The gelatinous form, 95 percent water, had been absorbed into the flats as the jelly became stranded by either an incoming or outgoing tide.

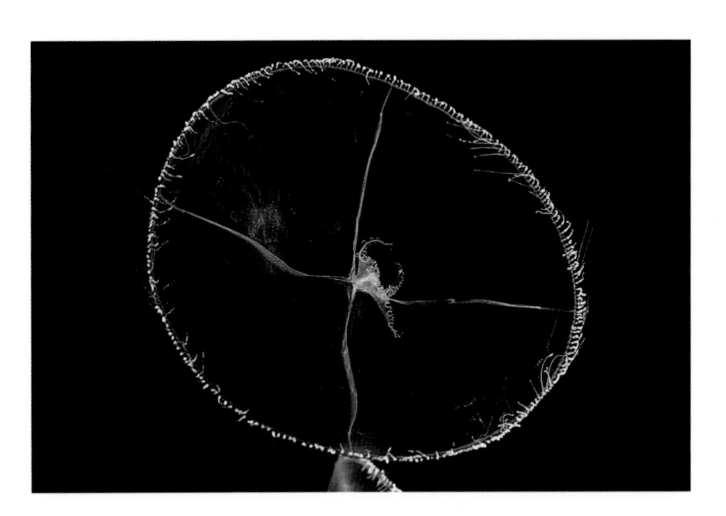

Mitrocoma cellularia, *the cross jelly, bears a cross of four radial canals.*

Jellies are wonder in motion, our imagination, our thoughts propelled. They can dance. They can sting. They can create a music not heard but felt in their staccato movements as they fly, pulse by pulse, through the ocean. Or they can simply rise in the sea as an idea, part of a much larger consciousness than their own. The Flemish painter Hieronymus Bosch places the transparent dome of a jellyfish in the center panel of his medieval triptych, "The Garden of Earthly Delights." Three individuals are talking inside. Call it a bubble of conversation, a dome of conversion. If we could hear what they were saying, an amplification of all they are discussing, considering, questioning, could we imagine alive on their tongues, "Above all the senses?"

Perhaps this is what moves me most about jellies—their sensory intelligence perceived through the clarity of their bodies, the great hunger that is sent outward through the feathery reach of their tentacles. Imagine the information sought and returned, communicated through the electrical circuitry of their nervous system: Where is the sun? Which way is the pull of gravity? What is food? Is there danger here? Predator or prey? Is it time to drift? Is it time to propel ahead?

I asked the biologist Ed Seidel who works with jellies at the Monterey Bay Aquarium what he had learned from them. He paused momentarily and then said, "It is heartening to know you can still be elegant without a brain reflex."

We are so proud of our gray matter as human beings, but we are such newcomers to the earth. Perhaps it is not enough to simply think our way through the world. More importantly, how do we feel it, sense it, absorb it? We have much to learn from these water-based creatures who so fully perceive the world around them in elegance and beauty.

Beauty is the beginning of terror, Rilke reminds us. What do we know? What will we ever know about the umwelt of jellies? Our limited knowledge even in the midst of science's remarkable claims fuels its own humility.

Part of my own terror as a *Homo sapien* resides in a preoccupation with my own species. What are we missing as a result of our anthropocentric orientation of walking upright on land with a large brain and two eyes focused ahead? What might we come to understand by imagining the world from a fluid point of view? Can we allow ourselves to be influenced by these pulsating, gelatinous forms that can throw out some semblance of a fishing net for corralling prey and reel it back in to their own transparent

Known originally from the Black Sea, the medusa, Maeotias inexspectata, *has taken hold in California wetlands.*

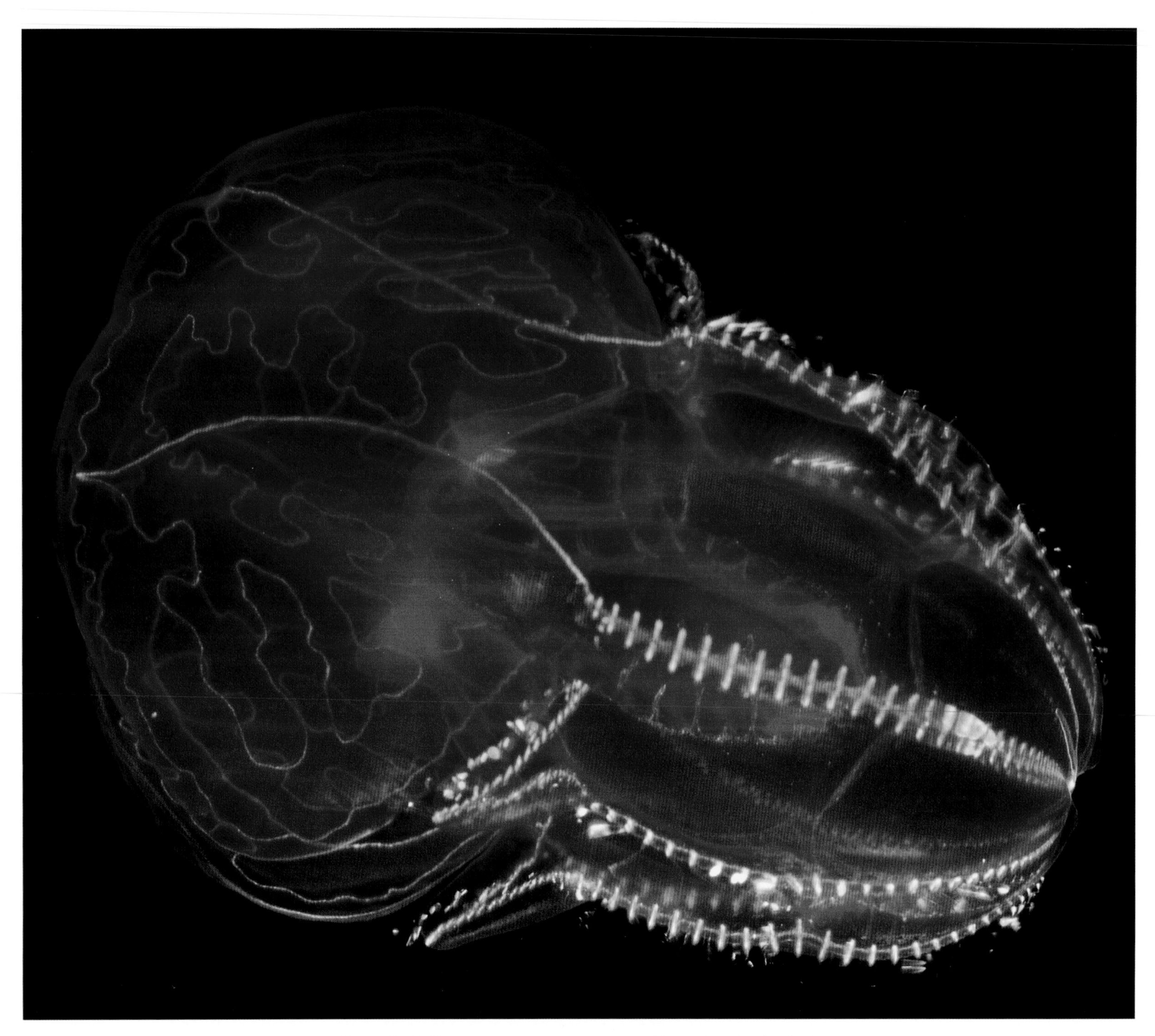

The comb jelly, Lampocteis cruentiventer, *resembles a ruby-colored walnut.*

bodies? Can we allow ourselves to be transported by the ciliated plates of comb jellies, their shimmering iridescence that flashes red, yellow, and blue, becoming tiny spaceships motoring through the pelagic cosmos, rising and falling with the currents? Can we ever fully penetrate their alien sensibility?

For me, it is enough to simply feel their remote, inaccessible, pulsing presence. Enough to acknowledge that the success of their free-floating reign on the planet's great seas is their persistence to keep moving, sensing, adapting, even in darkness, to the world around them.

Not long ago, my niece Diane and I went kayaking in the Elkhorn Slough near Monterey Bay. We were blessed to be in the company of Judith Connor and Nora Deans, the vibrant co-authors of this book, along with Michelle McKenzie and Miki Elizondo, the publication staff of the Monterey Bay Aquarium. We embarked on a journey to see wild jellies in the warmth of a midsummer day. We wanted to see outside what the aquarium has preserved on the inside.

Again, we took delight in the circle of life that surrounded us as we paddled in and out of the tidal creeks: harbor seals beached on the mudflats, sea lions and otters curious by our presence surfaced repeatedly, brown pelicans diving beak first were close enough for their entry splash to spray us with salt water, and the rookeries of great blue herons could be seen from afar in the screen of eucalyptus trees. And least terns, endangered, flew overhead, occasionally hovering for a better vantage point of what fish swam below. It was difficult not to feel both the luxury and loss of these coastal wetlands in California, where 75 percent have been destroyed by development and population pressures.

Judith and Michelle in a double kayak paddled back to show Diane, age eleven, an example of sea lettuce. Judith also explained the tidal ebb and flow of the slough, how the range between the highest tide and the lowest tide is around eight vertical feet. Worms and clams, crabs and snails, all adapt to a world submerged and a world exposed, twice daily.

Diane smiled, enamored with this watery world, all new to this Great Basin child of the desert.

We accompanied Nora and Miki, paddling into a side stream for a closer look at gulls. Their cries were deafening overhead.

Robinson Jeffers' words return to me:

Invisible gulls with human voices cry in the sea-cloud
"There is room, wild minds,
Up high in the cloud; the web and the feather remember
Three elements, but here
Is but one, and the webs and the feathers
Subduing but the one
Are the greater, with strength and to spare." You dream, wild criers,
The peace that all life
Dreams gluttonously . . . ah sacred hungers,
The conqueror's, the prophet's,
The lover's the hunger of the sea-beaks, slaves of the last peace,
Worshippers of oneness.

Diane and I returned to the main channel of the Slough, eager to negotiate the current. We were refining our paddle strokes, seeing how fast and straight we could navigate the kayak. The wind was at our backs and we were making excellent time. All at once, the color of the water changed. As if by magic or grace, we found ourselves in a school of sea nettles, *Chrysaora fuscescens,* as bright and luminous as suns with rays extended as oral arms and tentacles glowing several feet behind them. One after another streamed past us. With each orange-yellow pulsating umbrella, Diane cried out with delight, "Here's one, there's one—Look, over here, there, one just floated under our kayak—they're everywhere!"

We were brought into the haunting, exhilarating, mysterious relationship with Other. As human beings we could not comprehend the mind of these pulsating intelligences, we could only watch in awe as something much older and wiser propelled them in a direction we chose to follow. For a fleeting moment, we merged with a muscular, transcendent beauty whose inherent rhythm matched the beating of our own hearts.

A Gallery of SHAPE AND SIZE

I want to tell you the ocean knows this,

that life in its jewel boxes is endless as the sand,

impossible to count, pure.

—Pablo Neruda, from *Los enigmas*

by NORA DEANS

Walking along the beach, my toes sink into the sand just above the reach of the incoming tide. The warm, turquoise sea soothes me as the waves lap on shore in a calming rhythm.

A glint of something shiny catches my eye. I bend down to touch the transparent blue float of a tiny Portuguese man-of-war stranded on the sand, its long, purple tentacles stretching more than a meter. Still glistening from the sea, it no longer lives. This beautiful drifter washed in on the currents. I marvel at how these stunning creatures ride the winds like the kite

◂ *As large as a softball,* Crossota alba *lives deep in Monterey Canyon. Near it swims a small jelly predator, the nudibranch,* Phylliroe.

▴ *The hand-size floats of stranded Portuguese man-of-wars,* Physalia physalis, *litter a tropical beach and belie their elegance in the water.*

surfers racing just offshore, but how on land they become shapeless blobs of color. Theirs is a totally alien world, one without boundaries as we know them.

The wilted jelly on the beach is one animal, yet was once thought to be a colony of animals, each relying on the others. Jellies defy simple descriptions. In my mind, their bizarre shapes and sizes spring from the imaginations of artists and sculptors, not from the natural world. Translucent bells, glittering boxes, luminous chains, gossamer webs . . . these intricate, pulsing beings from a fairy-tale world dazzle and beguile.

From nearly microscopic tinkerbells to golden giants whose bells loom larger than beach umbrellas, you couldn't dream up more exotic jellies than already exist. Imagine animals that are 95 percent water, shrink when there's no food, drift freely on the currents, diffract prisms of light all over their bodies, live in a commune, eat their own kind, withstand the frigid, sunless depths of the deep, thrive in warm, freshwater bays, dance in the icy waters of the frigid poles and pack a deadly wallop with stinging cells that shoot out harpoons to snag other animals.

The mysterious life histories of these creatures elude us even today, despite all our technology for studying them in the deep. We're just discovering their real stories as we catch glimpses of these creatures from our dream world.

It's a world that shelters fluttering sea angels and delicate sea butterflies, prowling lion's manes and tinkling bell jellies, glittering crystal jellies and sinuous salps, sparkling sea slugs and racy by-the-wind sailors. A world of cannonball jellies and firework siphonophores, of stinging sea nettles and fearsome man-of-war. So many startling shapes and sizes, a living gallery of original art that pulses and glows and shifts and vanishes, a gallery in motion.

▲ *A giant in the jelly world, the bell of the golden lion's mane,* Cyanea capillata, *could envelop a large beach ball.*

Every moment something new appears in this gallery. Long, looping chains called siphonophores pass by, nearly 40 meters long, dwarfing even blue whales. These communities live as one, with living fishnets, swimming bells, floats, stomachs and reproductive organs all jumbled together for survival.

Nearby, lacy larvaceans spin sticky webs of translucent jelly around their little bodies. They live tucked inside until the wear-and-tear of life in the currents shreds their delicate homes, and they cast them off to weave

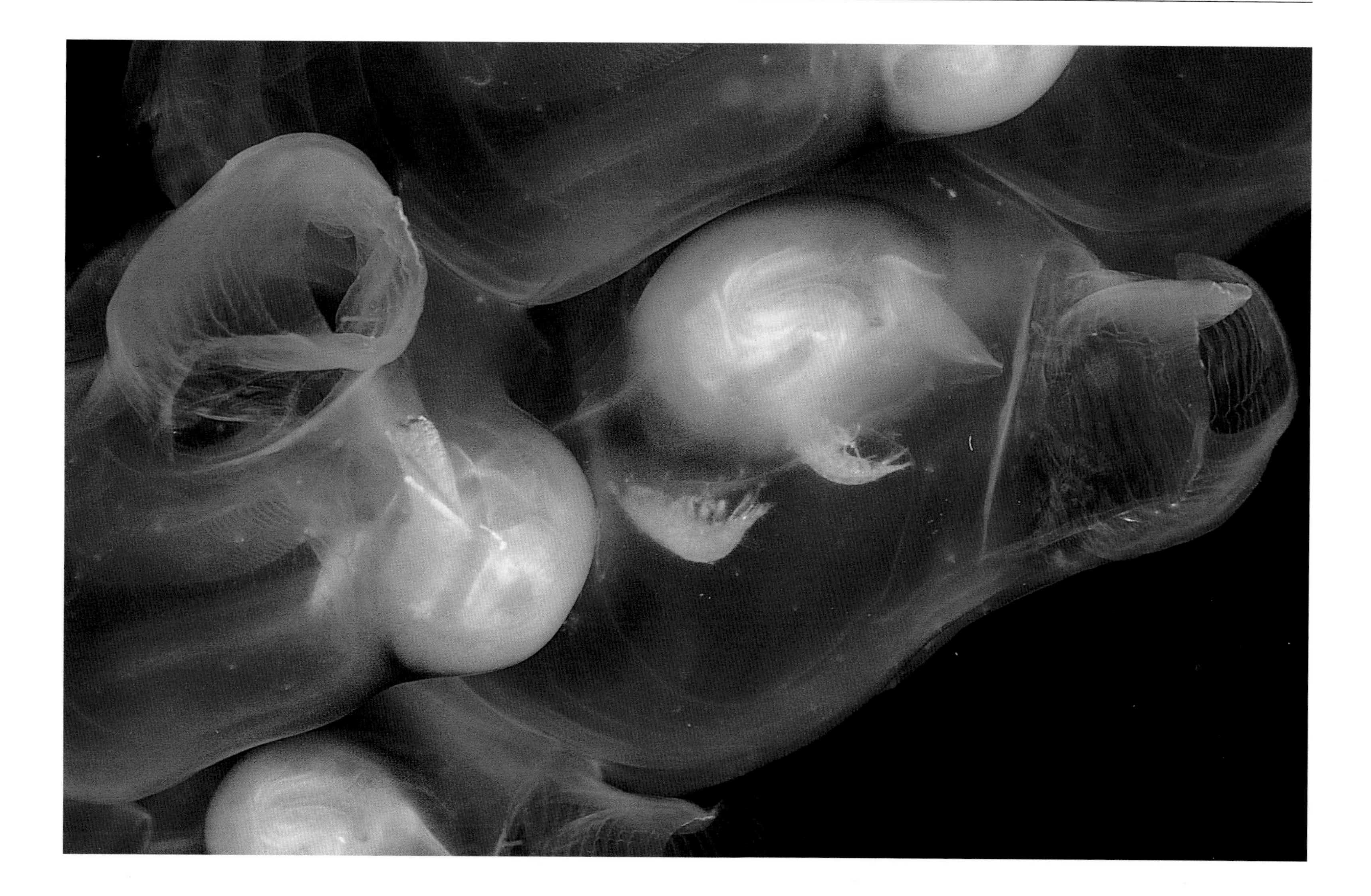

new ones. The tattered webs sink into the depths, to be fed upon by the thousands drifting below.

Wait a moment and you may spy barrel-shaped salps, transparent as finely spun sugar. Or a tomato-red jelly lying on the bottom, its harpoon-studded tentacles fishing the sea around it. A luminous moon jelly pulses in the shallows, as if fallen from the night sky. It's a world of living art, where

◄ *The red gut deep within the gossamer net of this tiny lobed comb jelly,* Bathocyroe fosteri, *hides any bioluminescent prey the comb jelly consumes in the lightless ocean depths it calls home.*

▲ *Small pink hyperiid amphipods hitchhike inside a salp, using it for protection and perhaps as a tasty snack.*

▲ *Resembling a delicate pastry, fluffy yellow gonads surround the red mouth of the little bottom-dwelling jelly,* Ptychogastria polaris.

frilly mouth-arms embellish the most simple of shapes, giving rise to "visions of floating lingerie," as my friend, Connor, would say. Her passion for seaweeds inspires others to come closer, to explore this riotous world of color and motion.

In this dynamic gallery, the art packs a mean punch. Jellies and their kin sweep the seas like millions of drift nets, catching hordes of tiny plankton or fish that veer into their path. Rachel Carson wrote of the vast reach of jellies in *Under the Sea Wind:* "That night as a school of young mackerel swam

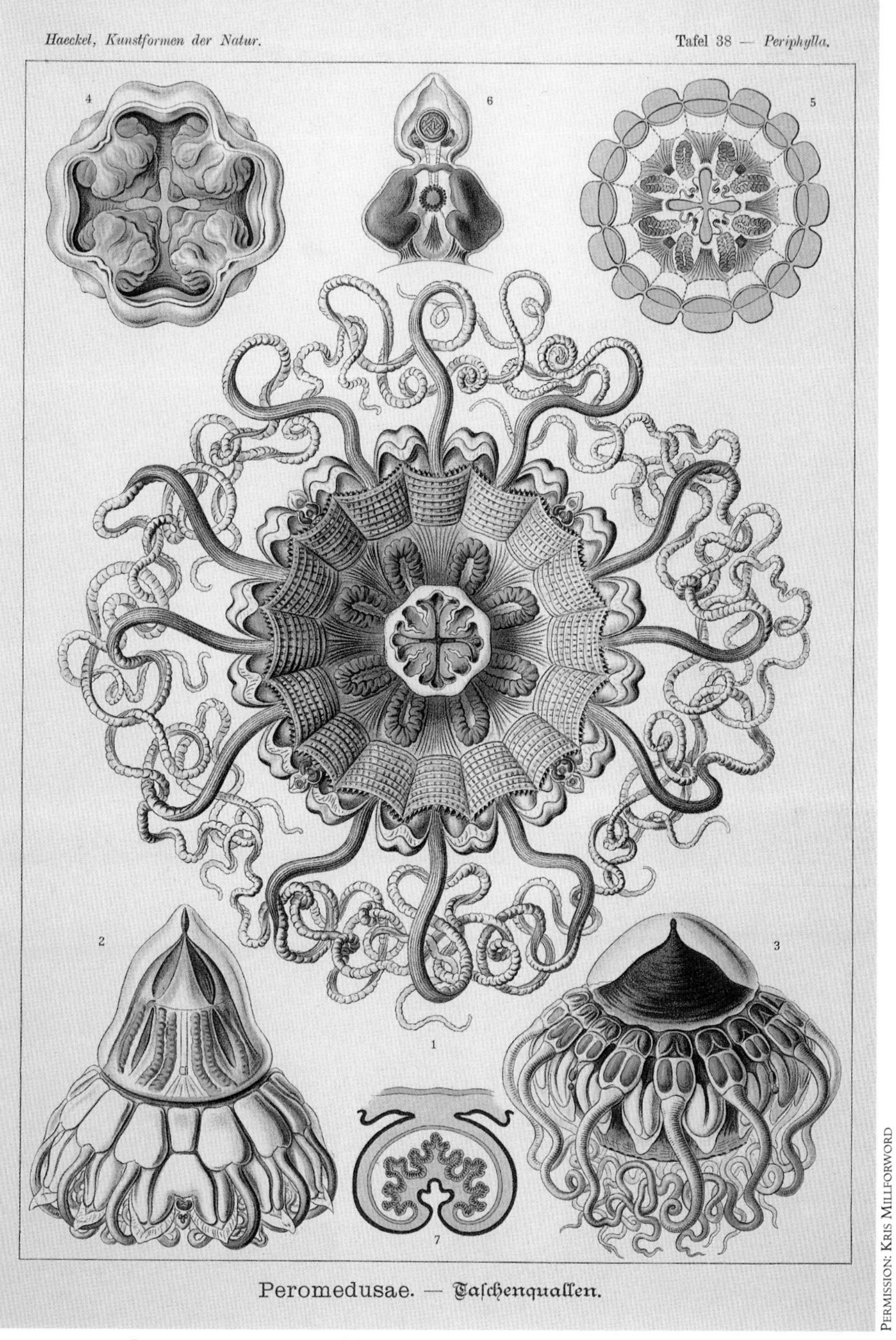

Illustrations from The Medusae, 1909 ■ Ernst Haeckel

▲ *Ernst Haeckel with his assistant, Nikolaus Miclucho-Maclay, 1868.*

◄ *Illustration of jelly* Periphylla

Using an artist's eye for line, space and color, and a scientist's attention to detail, scientific illustrators portray the natural world with beauty and precision.

For hundreds of years, illustrators have helped scientists document animal and plant species. Working from live organisms, preserved specimens and written and personal accounts, they create accurate visual records that inform and delight.

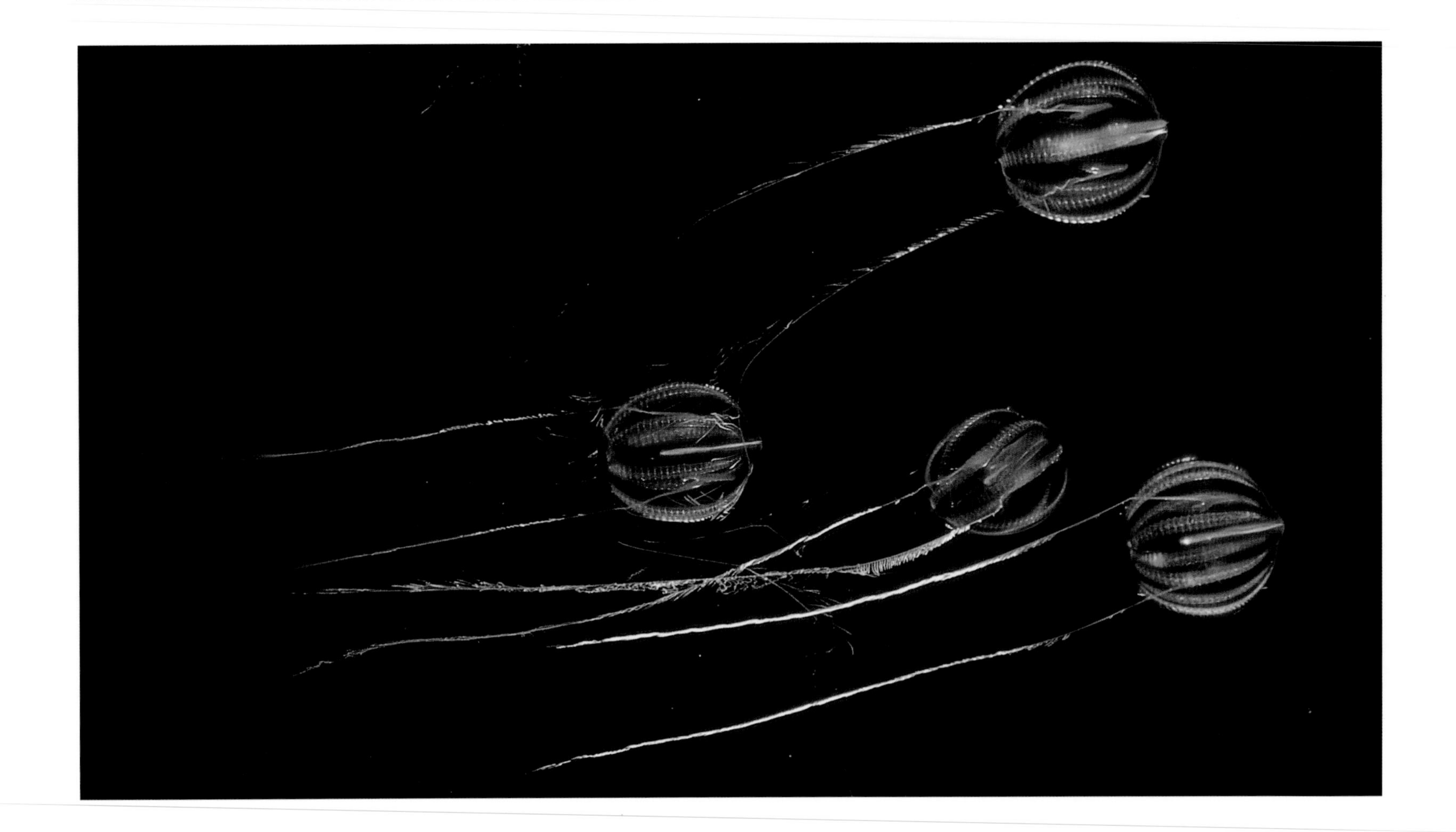

▲ *Trailing shimmering tentacles, tiny sea gooseberries,* Pleurobrachia bachei, *spin and twirl to catch other small sea creatures in their path.*

▶ *More than twice as large as sea gooseberries, this lobed comb jelly,* Deiopea kaloktenota, *swims near the surface of Monterey Bay.*

near the surface, they passed over a sea of death, for ten fathoms below them lay millions of the comb jellies, *Pleurobrachia,* in layer after layer, their bodies almost touching one another, twirling, quivering, tentacles extended and sweeping the waters as far as they could reach, sweeping the water clean of every small living thing."

I scan the turquoise waters and the sun-baked beach ahead, but see only swimmers and surfers. As I walk on, I think about this other world, where Portuguese man-of-war swarm by the thousands, haunting tropical waters

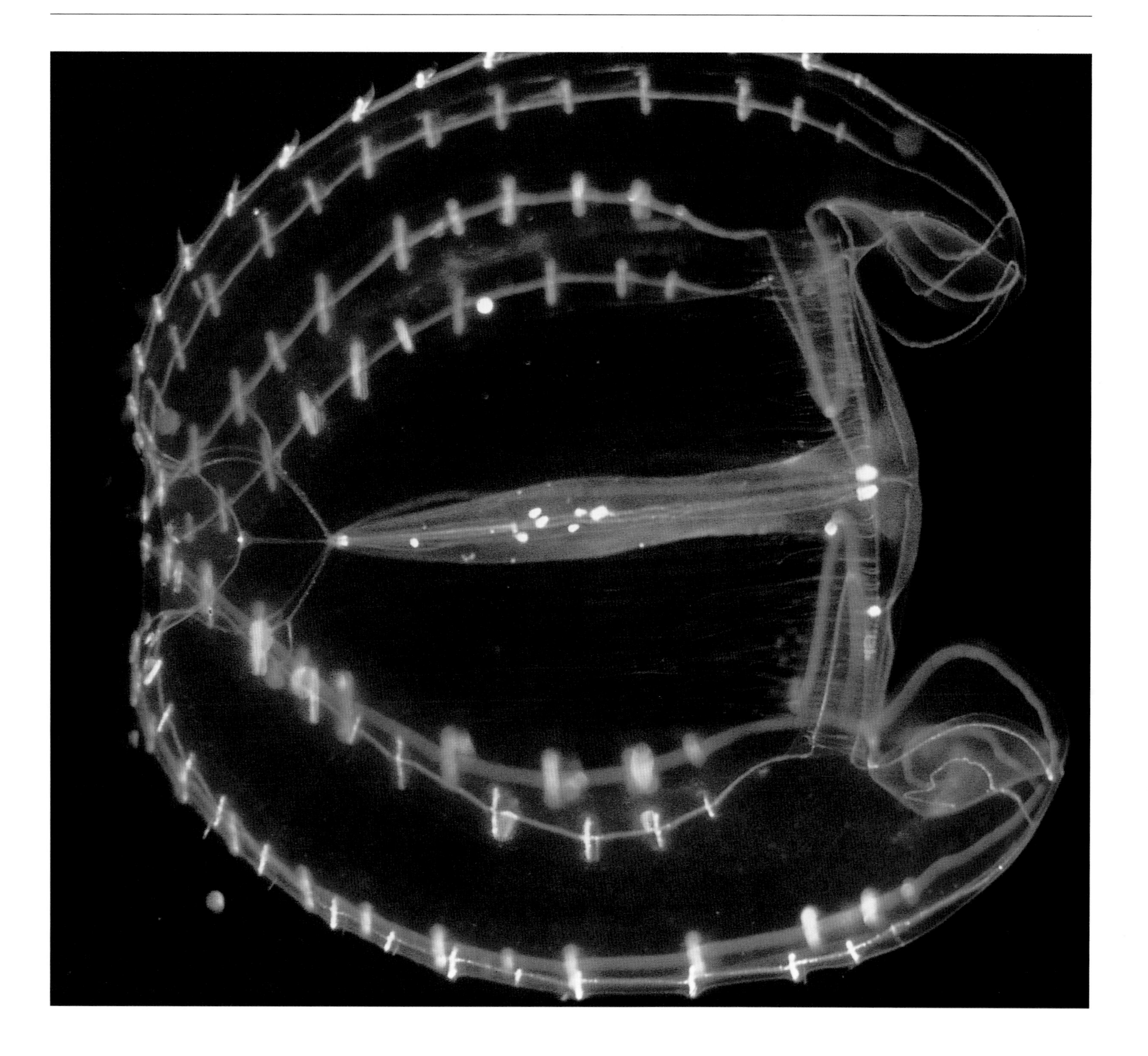

here in Hawaii and elsewhere, striking fear in swimmers who dread their nasty stings. Long tentacles trail more than 46 meters, snaring those who never even see the purple float.

Back in my bungalow, I look up the Hawaiian names of these creatures, *Ili Manéo, Páimalau Palalia* or *Poloia,*—names given to any jelly-like creature. Marine scientists rarely call them 'jellyfish' since they're really not fish at all. They lack important characteristics of fishes, like bony skeletons, brains or hearts.

Since Aristotle's day, people have grouped animals by things they share, or things they lack. All animals that carried harpoon-like weapons called "cnides," (Greek for 'nettle') ended up in the same clan called Cnidarians. These include the true jellies, like sea nettles and moon jellies, as well as box jellies and siphonophores, and their cousins, corals and sea anemones. And that's a clue to one of the mysterious life stories many true jellies share.

◂◂ *Northern sea nettle,* Chrysaora melanaster

◂▴ *Close up of the business side of an upside-down jelly,* Cassiopeia xamachana

▴ *Portuguese man-of-war,* Physalia physalis

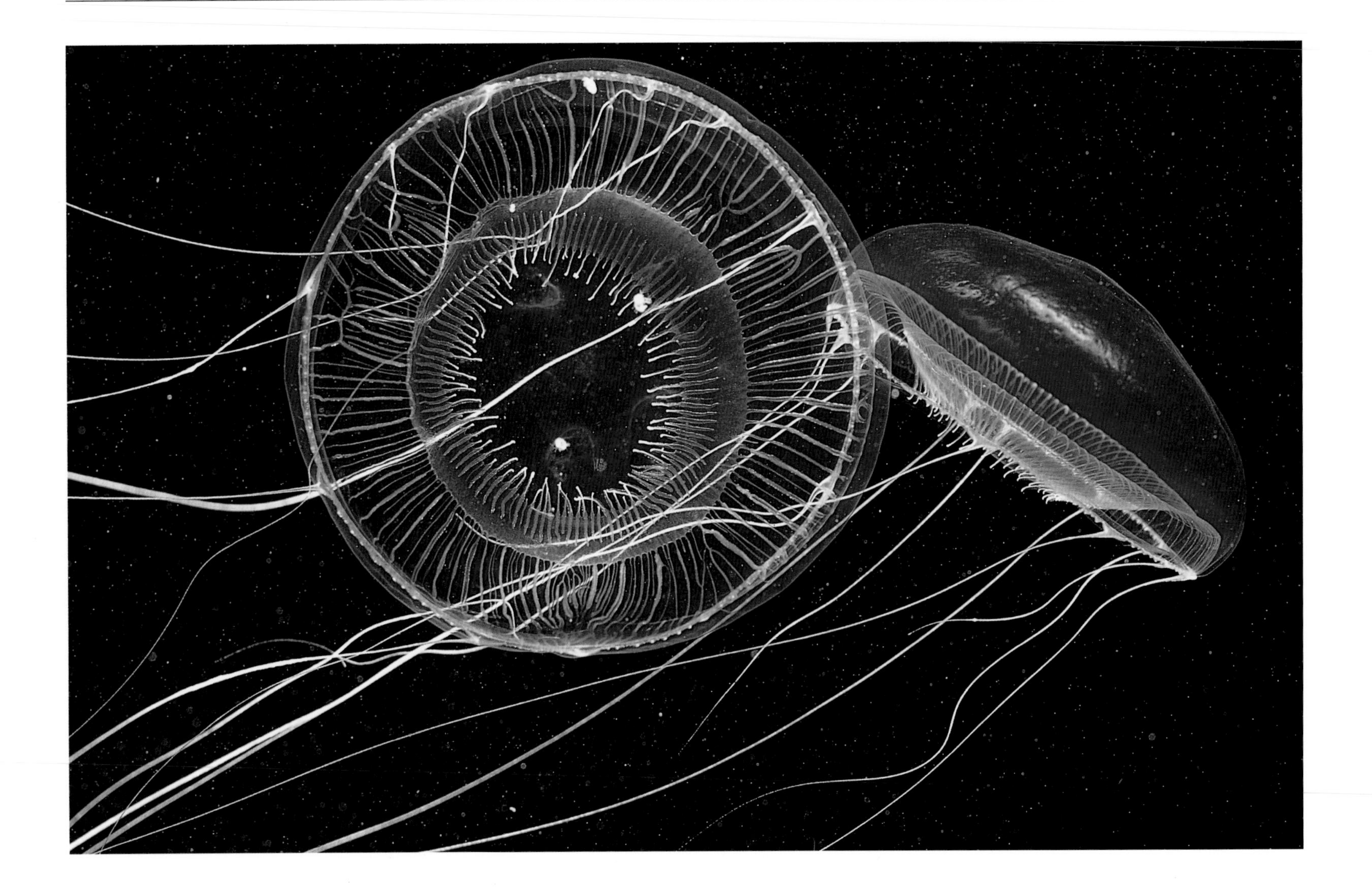

▲ *Crystal jellies,* Aequorea *sp., display trailing tentacles recalling why the body plan of the medusae was named for the snake-headed Medusa of Greek mythology.*

They're not always pulsing, umbrella-shaped bells with thread-like tentacles and ruffled mouth-arms. Some spend part of their lives as polyps living attached to something, upside down. Entire generations. And others don't have tentacles at all.

It's a plan that's worked for more than 700 million years. Since before the dinosaurs. And it seems the more we change the seas today, by overfishing or

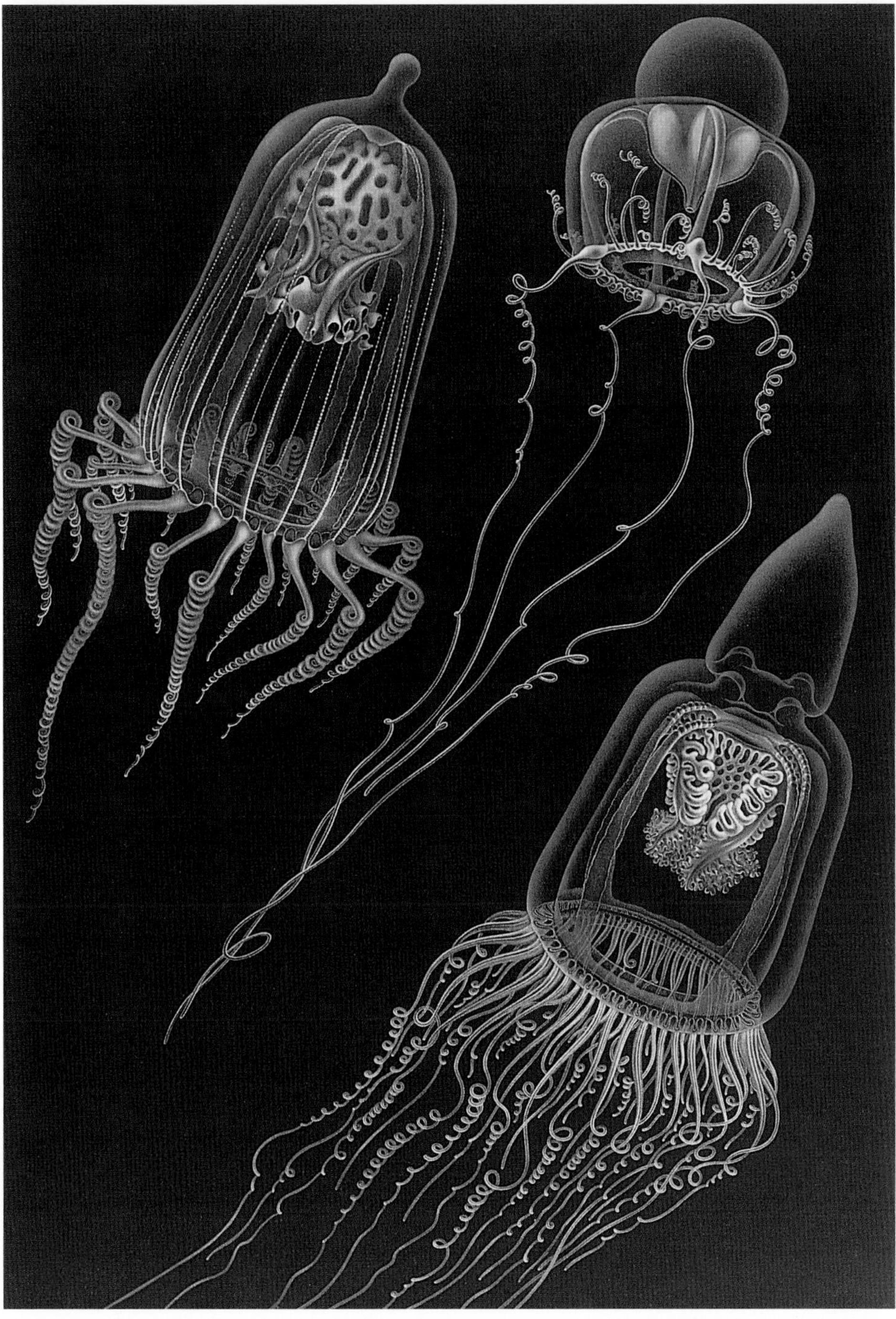

Cnidaria of the Mediterranean, 1970

Ilona Richter

Following a time-honored tradition, Richter's scientific illustrations depict jellies and other animals in exquisite, accurate detail.

As you look at these illustrations, notice how the use of color captures the jellies' transparency—you can both see them and see through them. And notice how the use of simple lines can define tentacles, from thick spirals to hair-fine threads.

◂ *Scientific illustration of jellies by Ilona Richter, from* Monograph 39 *of the* Fauna and Flora *series.*

▲ *Nearly transparent except for a small, spiral shell, a heteropod sea snail,* Carinaria cristata, *rapidly flexes its odd-shaped body to escape.*

▲▲ *The light from a pyrosome can be seen for up to 100 meters!*

▶ *The tentacles of a black sea nettle,* Chrysaora achlyos, *shelter these juvenile fish who are somehow immune to their stings.*

altering habitats, or introducing species to new waters, the more jellies thrive, swarming in some seas to the point of clogging fishing nets and intake pipes. That puzzles me. How do these fragile-looking creatures survive in polluted waters? Not all do. But when run-off from land carries fertilizers into the coastal waters, it feeds the tiny plant plankton that support the ocean food chain. Jellies and others feast on the resulting bloom, and prosper. And when we catch tons of fish from areas like the Bering Sea, we may give the upper hand to jellies who have fewer fish to compete with for food.

I think about that shriveled jelly stranded on the beach, and how truly its kin are creatures of the watery world we can only visit. A soft body like that would never survive in our dry land of hard-edged boundaries. John Steinbeck and Edward Ricketts described their futile efforts to collect jellies in *The Sea of Cortez:* "About noon we moved through a great group of Zeppelin-shaped jellyfish, ctenophores or possibly siphonophores. They were six to ten inches long, and the sea was littered with them. We slowed down and tried to scoop them up, but the tension of their bodies was not sufficient to hold them together out of water. They broke up and slithered in pieces through the dip nets."

The buoyant lifestyle of a jelly lets it reach huge proportions. The giant black sea nettle *Chrysaora achlyos,* a classic true jelly, pulses its large, tan-flecked, maroon bell in calm waters offshore, away from the damaging surf zone. Twenty-four tentacles dangle seven meters or more. Black sea nettles have become much more common off the coast of San Diego, feasting on local plankton blooms.

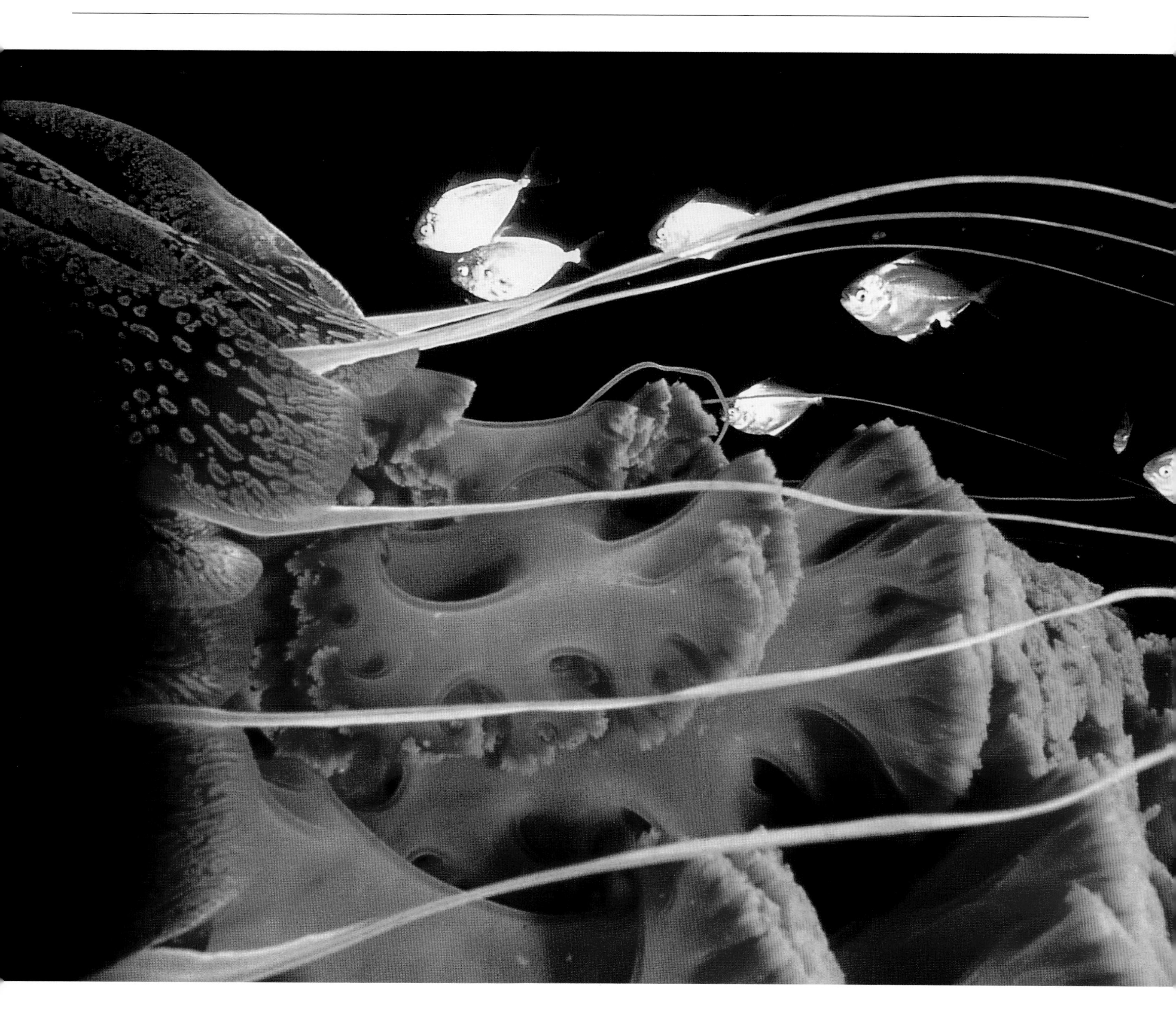

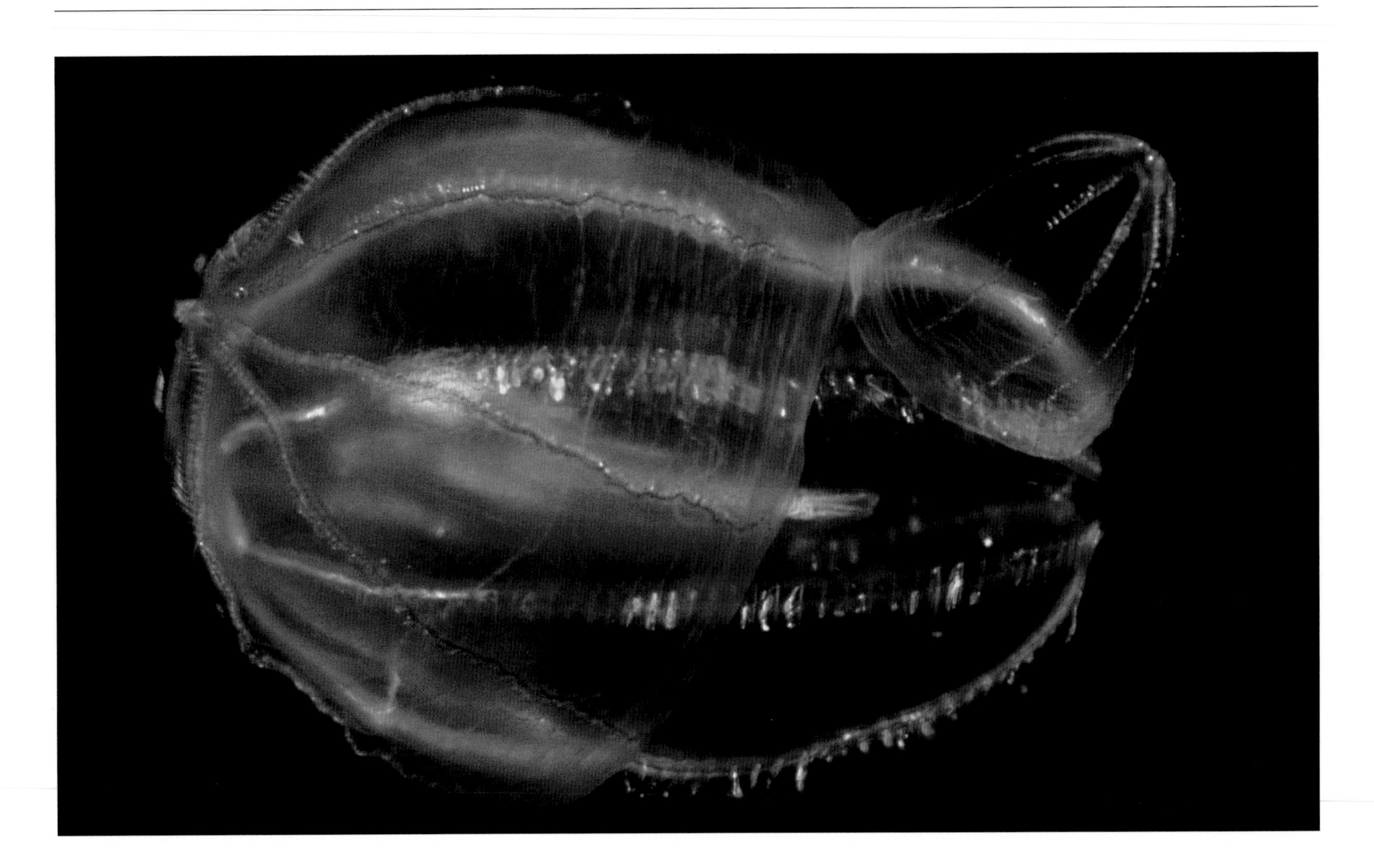

▲ *It's often a jelly-eat-jelly world, even if the other jelly is larger. In this case, two comb jellies,* Beroë gracillis, *are eating a sea gooseberry,* Hormiphora.

▶ *Tentacles from the black sea nettle,* Chrysaora achlyos, *trail for 8 meters or more, fishing the waters as the nettle pulses through coastal waters.*

While the black sea nettle fits our image of jellies, other jelly-like creatures test our sense of the bizarre. Some comb jellies, or ctenophores, don't have bells or tentacles. Like swimming mouths, they zip around in the sea, devouring tiny plankton and sometimes each other, diffracting rainbows of light.

I'm captivated by these jelly creatures and their sheer diversity. I want to know more. Why can I see through some, while others vibrate with color?

▲ *Virtually transparent, a crystal jelly,* Aequorea victoria, *is hidden even out in the open.*

What do the different stripes and dots and patterns mean? What can I tell about their lifestyle by their shape and motion?

I call Connor as soon as I get home from Hawaii. "Remember how we talked of writing another book together? What about jellies?"

We meet again in the two places on the cutting edge of searching for the answers to these questions, the Monterey Bay Aquarium Research Institute, and her sister, the Monterey Bay Aquarium. The aquarium creates hauntingly captivating exhibitions of jellies. One of my favorite exhibits showcases jellies as living art, surrounded by truly wondrous works of amazing artists. The natural history artists draw me closest. Bigelow's *Illustrations from the Medusae.* The Blaschkas' *Three Glass Jellies.* Ilona Richter's *Cnidaria of the Mediterranean.*

For years I've collected prints of fishes, plants, birds and animals, old and new. The attention to detail, the painstaking accuracy and beauty captured on paper, represents hours and hours of looking and observing. Hours spent in the field, sketching, watching, waiting. Look at the incredibly detailed work of Leopold and Rudolf Blaschka, father and son, who created thousands of glass specimens, accurate to the smallest detail. I read the exhibit label, "So magically beautiful and detailed was their work, it was described as an artistic marvel in the field of science and a scientific marvel in the field of art."

To really know nature, draw it, paint it, write about it. And then perhaps you'll care enough to try to save it. In some cases, these masterpieces of illustration are all we have of long-lost species. I want to turn that tide. I want people to know, and understand and care as much as these artists cared. Their passion reaches out to us. In David Hockney's words, "I believe that art can change the world."

PHOTO: FRANK J. BORKOWSKI

▲ *Blaschka jellies, Cornell University.*

THREE GLASS JELLIES, CIRCA 1885 ■ LEOPOLD AND RUDOLF BLASCHKA

Using simple tools, glass rods, wire, glue, paint and paper, this father-and-son team created scientific models that were both stunning and true-to-life.

Notice the graceful curves of the jellies' bells, their depth, transparency and delicate coloring. See how supple they look, as if drifting sensuously in water, not frozen in air.

Dancing Water
RHYTHM AND MOTION

During much of the summer they have been drifting offshore,

white gleams in the water.

sometimes assembling in hundreds along the line

of meeting of two currents . . .

—Rachel Carson, from *The Edge of the Sea*

by JUDITH CONNOR

Blue light surrounds us as we stand before the exhibit of pulsing jellies and listen to the comments of other visitors. The hall is abuzz with conversation, but as families enter the dimly lit gallery, all chatter stops. A hush falls over us. I turn to my friend, Nora, who nods her head in understanding. By the sea nettle display a woman whispers to her friend, "They're so beautiful . . . almost transparent." The twinkling lights of comb jellies

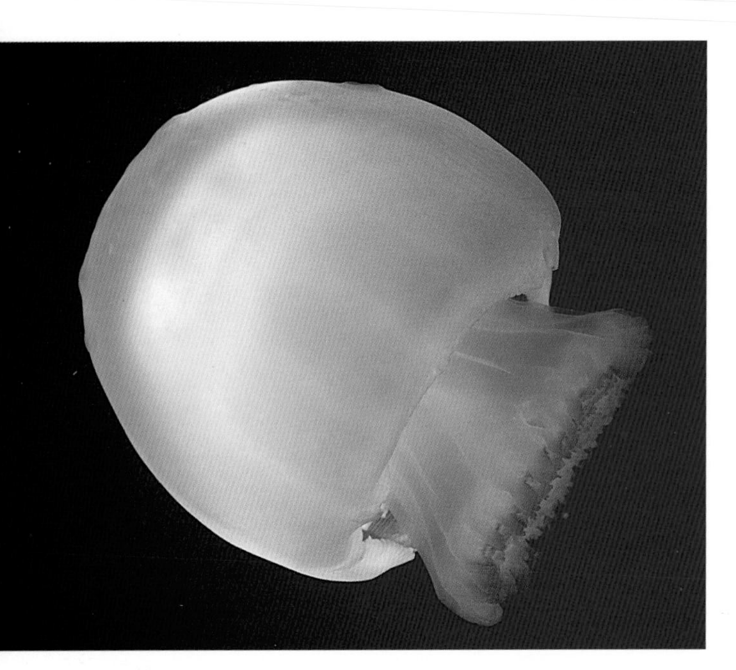

◂ *The sea nettle* Chrysaora fuscescens, *glides continuously if slowly, trailing slender tentacles and ruffled oral arms in the current.*

▴ *Pulsing its thick rigid bell, the cannonball jelly,* Stomolophus meleagris, *simulates a lively tempo.*

▸ *The by-the-wind sailor,* Velella velella, *drift on the sea surface with triangular sails held aloft to catch the breeze.*

captivate two young boys. A child stands on tiptoe and points to the smallest ones, "Look, Mom, they're baby jellies."

"I want to see the lithograph of David Hockney's watery pool." I tell Nora, "I once drove 500 miles to see his paintings, amazed to think that with just a few lines he can catch a splash in midair." Nora reads the caption below the picture, "Like the still time between deep breaths, Hockney's work encourages us to simply be—even for just a few seconds—in the calm stillness of the present moment."

Later, in a dark video laboratory at the Monterey Bay Aquarium Research Institute, that sense of calm returns. We watch jellies in the spotlight of the remotely operated vehicle's (ROV) camera, and talk of art and the physics of jelly locomotion. Watching jellies that few have ever seen, we're grateful for this chance to stretch the normal perimeters of our individual work.

Our view into the deep sea is framed by the ROV, *Ventana*. Worthy of her name, *Ventana* provides us a window into the watery niche that jellies inhabit. The submersible's pilots maneuver the ROV in close to the subjects of our delight, closer than we imagined possible. High-definition video cameras reveal the details of animal structure and locomotion. From the ROV's first splash into sunlit waters through its descent to the inky depths, we are spectators at an undersea ballet of drifting creatures.

On the sea surface, cobalt-blue hydroid colonies called by-the-wind sailors drift by. With triangular sails held aloft above their gas-filled floats, their sea voyages are driven by the wind. Some have sails set for westward travel; others from the same population are mirror images with sails set toward East. Wind patterns separate the two groups of travelers. Some sail

▲ *The spotted jelly,* Mastigias papua, *wanders its tropical lake home, following the sun by day and sinking to the lake floor by night.*

thousands of kilometers across the open ocean, only to be cast ashore on beaches up and down the coasts they encounter at journey's end. I muse on cobalt-blue fragments in beach wrack. Surely, they too could find a place in art—perhaps someday joining the shell fragments, bits of bone and swirls of watery color in a Pegan Brooke painting.

Motion on the screen draws my attention: sea nettles glide into view, adrift on ocean currents. Composed almost entirely of water, these jellies are truly one with their environment. They float easily through the water, trailing long tentacles that surround lacy mouth-arms. Some have escorts, like the small fishes that find sanctuary among the sea nettles' tentacles, doing the fish no harm. The casual drifting of the sea nettles is punctuated by slow, rhythmic pulsing of their bells, a form of jet propulsion. Contracting a thin layer of muscles, the bell's margin moves inward. The movement expels a gentle jet of water and propels the animal forward. Then the elastic bell springs back to its normal size and shape as the muscles relax.

The tempo isn't a standard that's set for every kind of jelly. The shape and volume of each species' bell sets its speed. A relaxed mode of transportation expends less energy and requires less food than that of more energetic animals. While sea nettles glide slowly, the rounded shapes of cannonball jellies and spotted jellies advance with a lively staccato beat.

Five hundred meters below the sea's surface, the deep-sea jelly *Colobonema* swims like a sprinter. Its crystalline bell, nearly as deep as it is wide, confines a generous volume of seawater. In active flight, it contracts its bell margin in rapid sequence to slurp in and then expel the watery holdings. Quick pulse, pulse, pulse, pulse, it sprints with dangling tentacles of different lengths that

FROM THE COLLECTION OF AMY NATHAN AND WARREN WEBER. PHOTO: MARTY KNAPP

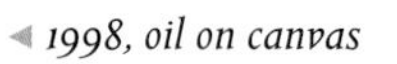

◄ *1998, oil on canvas*

DUXBURY REEF, 1 ■ PEGAN BROOKE

Watch for watery whorls of mingling hues, looping lines and pod-like shapes; find fragmented shells and bits of bone. Brooke's art speaks of the rhythms of the sea and of our connection to all living things.

► *Pegan Brooke in her studio, 2000*

PHOTO: MARY SEIDMAN

▲ *This colorful, cosmopolitan jelly,* Atolla vanhoeffeni *swims in short fast bursts through the midwaters more than 500 meters deep.*

▶ *Sixteen dark streaks of color mark the pale bell of the northern sea nettle,* Chrysaora melanaster.

bounce in cadence. Then, with a pause, the crystal jelly relaxes, slowly spreads its bioluminescent tentacles and drifts like a star in the darkness of the deep.

The vibrant beauty *Atolla* is not as fast, but with short bursts of speed, it can escape some predators. The cone-shaped *Periphylla* and saucer-shaped

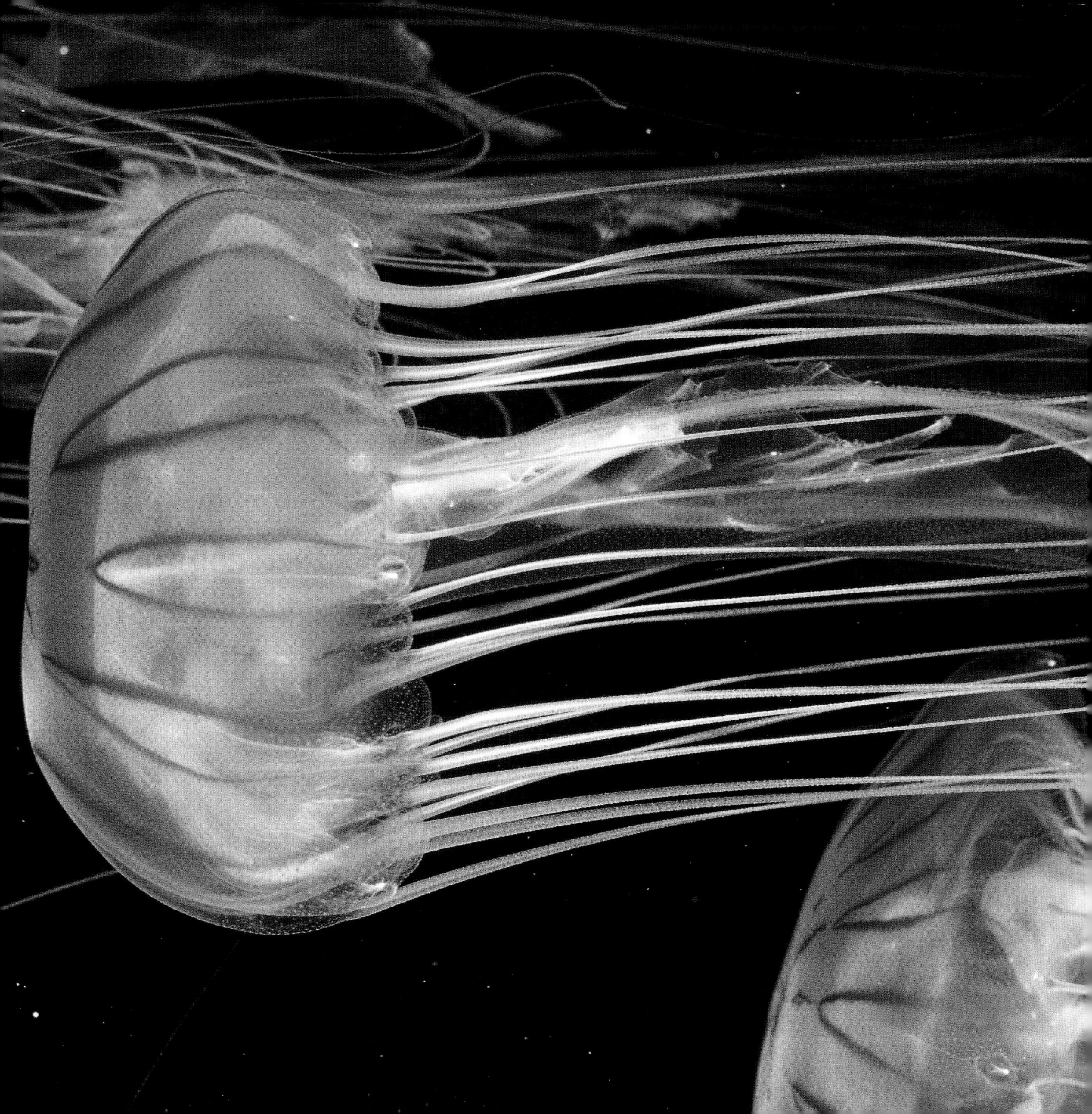

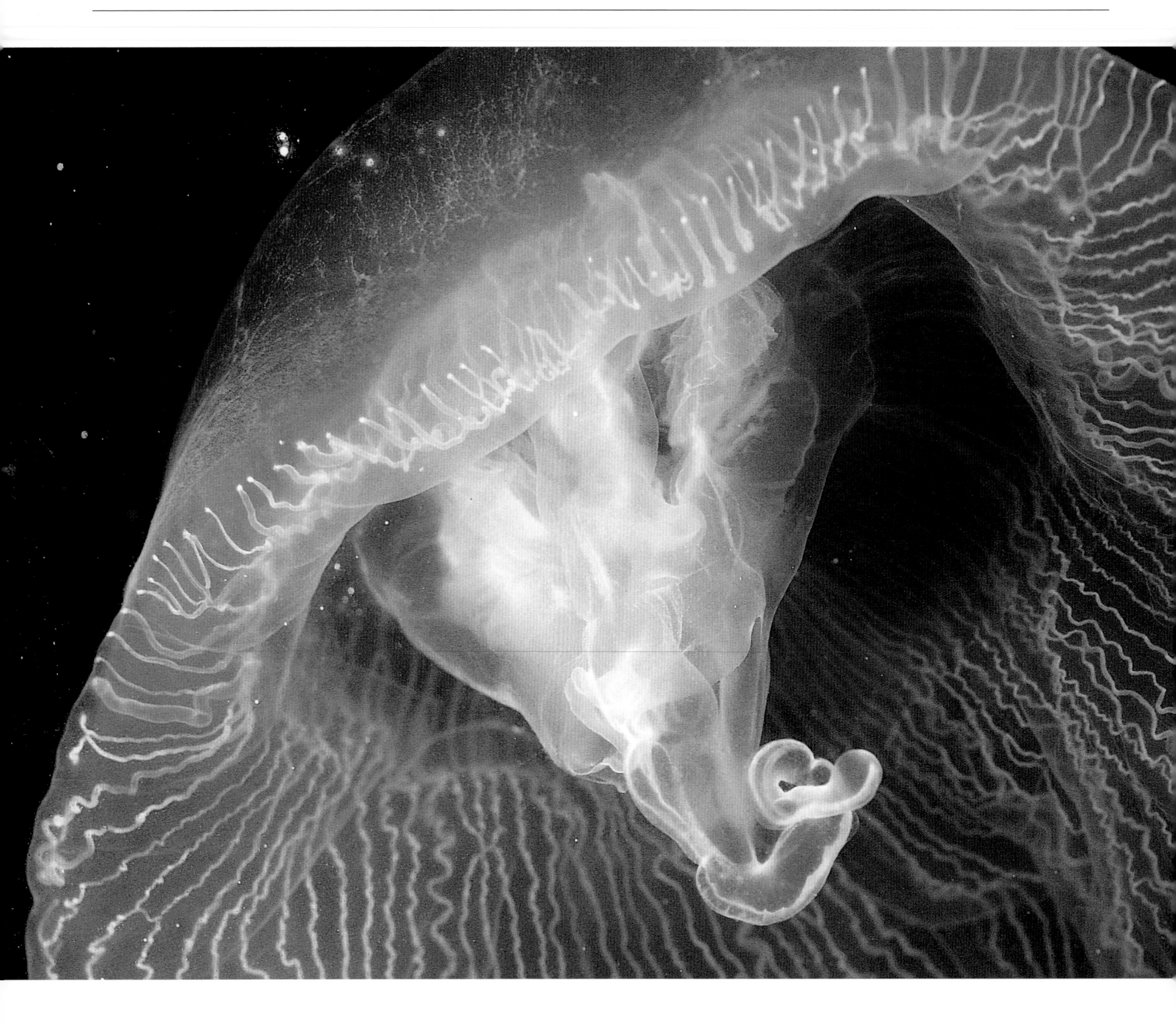

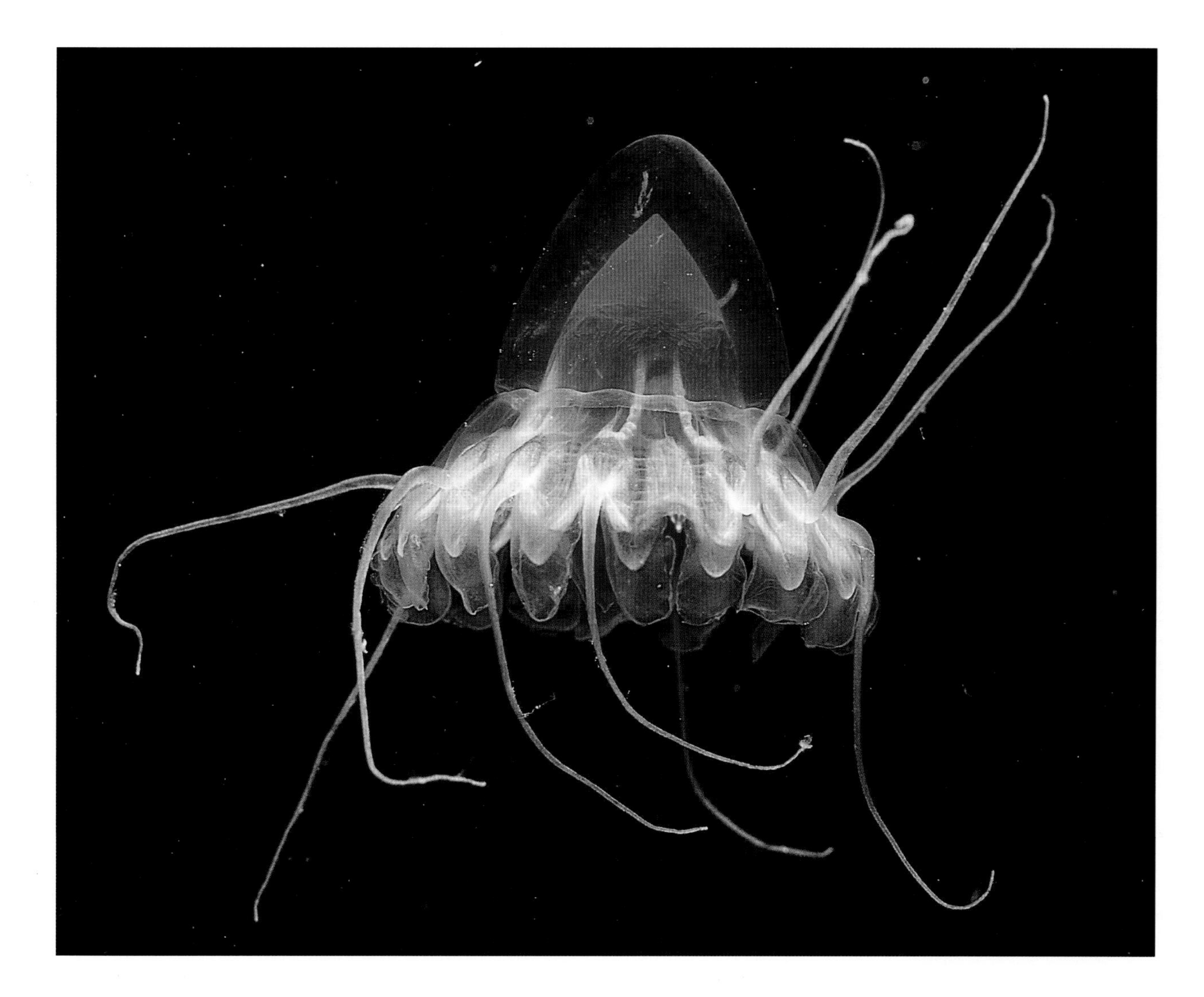

Poralia move even slower. But *Solmissus,* broader than tall, is a master of slow motion in the guise of an elegant crystal platter. This jelly's shallow bowl accommodates little of the seawater propellant that drives its gentle jets. A velum ribbon around the saucer's rim pulses, propelling *Solmissus* at a leisurely pace with its tentacles curved up and out. It glides and hovers like the flying saucer it resembles as it crosses our camera's path. Slower still is

◂ *The rare* Deepstaria enigmatica *is large and flimsy, swimming with slow peristaltic movements.*

▴ *Deeply colored and regally beautiful,* Periphylla periphylla, *migrates vertically with its tentacles held high.*

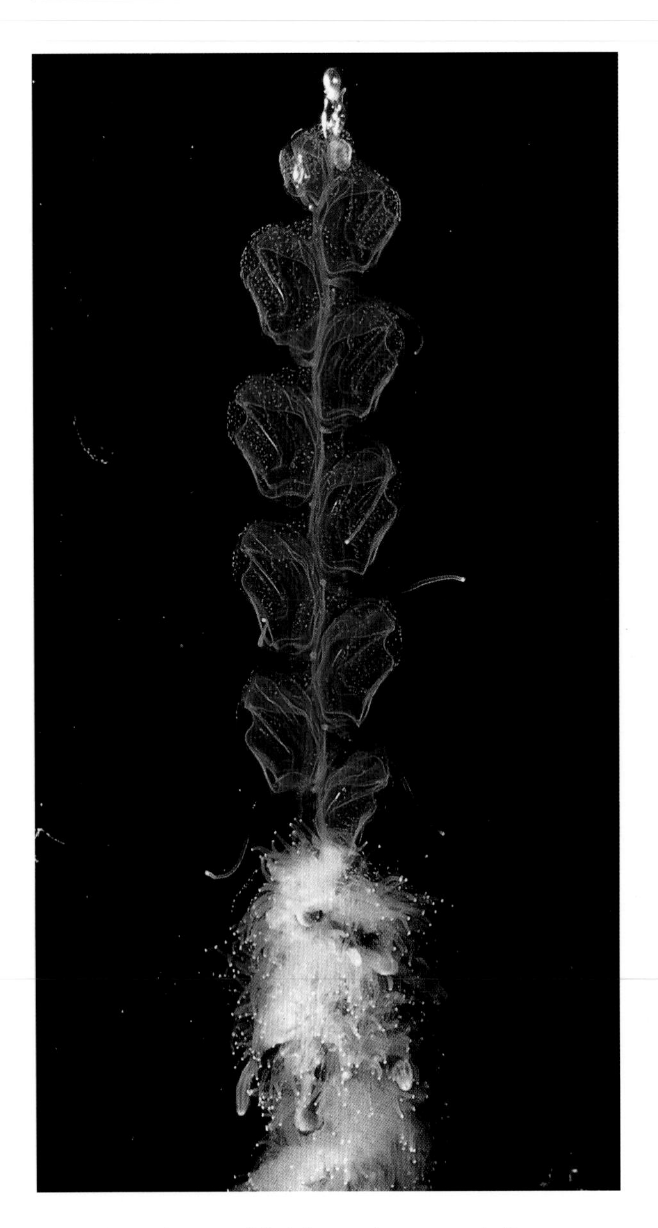

▲ *Leading with its gas-filled float and swimming bells, the long-stemmed beauty,* Apolemia uvaria, *trails its long gelatinous chain through the deep midwaters.*

▶ *The delicate siphonophore,* Lensia, *with its rocket-shaped swimming bell, darts with quick jets through the water.*

the enormous *Deepstaria,* which contracts its flimsy bell from top to bottom with quivering, peristaltic waves.

Special tools for deep sea study unveil to us the secrets of siphonophores. Using the ROV and sonar, we've measured siphonophores stretching out some 30 meters or longer in length. Compared to the Portuguese man-of-war at the sea surface, deeper-dwelling siphonophores have smaller gas-filled floats or swimming bells—or both. Our eyes and the camera navigate the length of *Apolemia* in search of its small float that absorbs and releases gas to regulate buoyancy. Just below the gas float, swimming bells pulse rhythmically, towing the siphonophore's long body chain behind. In sharp contrast, a rocketship siphonophore—just a few centimeters long—zips by. It has a rocket-shaped swimming bell, but lacks the gas float. Instead of using gas for buoyancy, it employs rapid pulses of its powerful muscles for jet propulsion to dart like a miniature missile. Before our eyes, the living rocketship contracts its trailing portion close to the swimming bell for streamlined speed. Then drifting in a relaxed mode, it splays its tentacles in a precise array for fishing.

Surely the most compelling of the animals we view are the ctenophores, or comb jellies. Irresistibly beautiful in motion, each bears eight comb rows with hundreds of comb plates that paddle water from front to back, or back to front. We watch a comb jelly propel itself, flashing iridescent light, rowing those paddles in a perfect succession of power strokes and recoveries. Light that strikes delicate, parallel lines on each comb plate is diffracted into those flashing rainbows. Pulsing colored light outward, the combs propel the animal forward. We spy on comb jellies that creep or fly, or undulate their

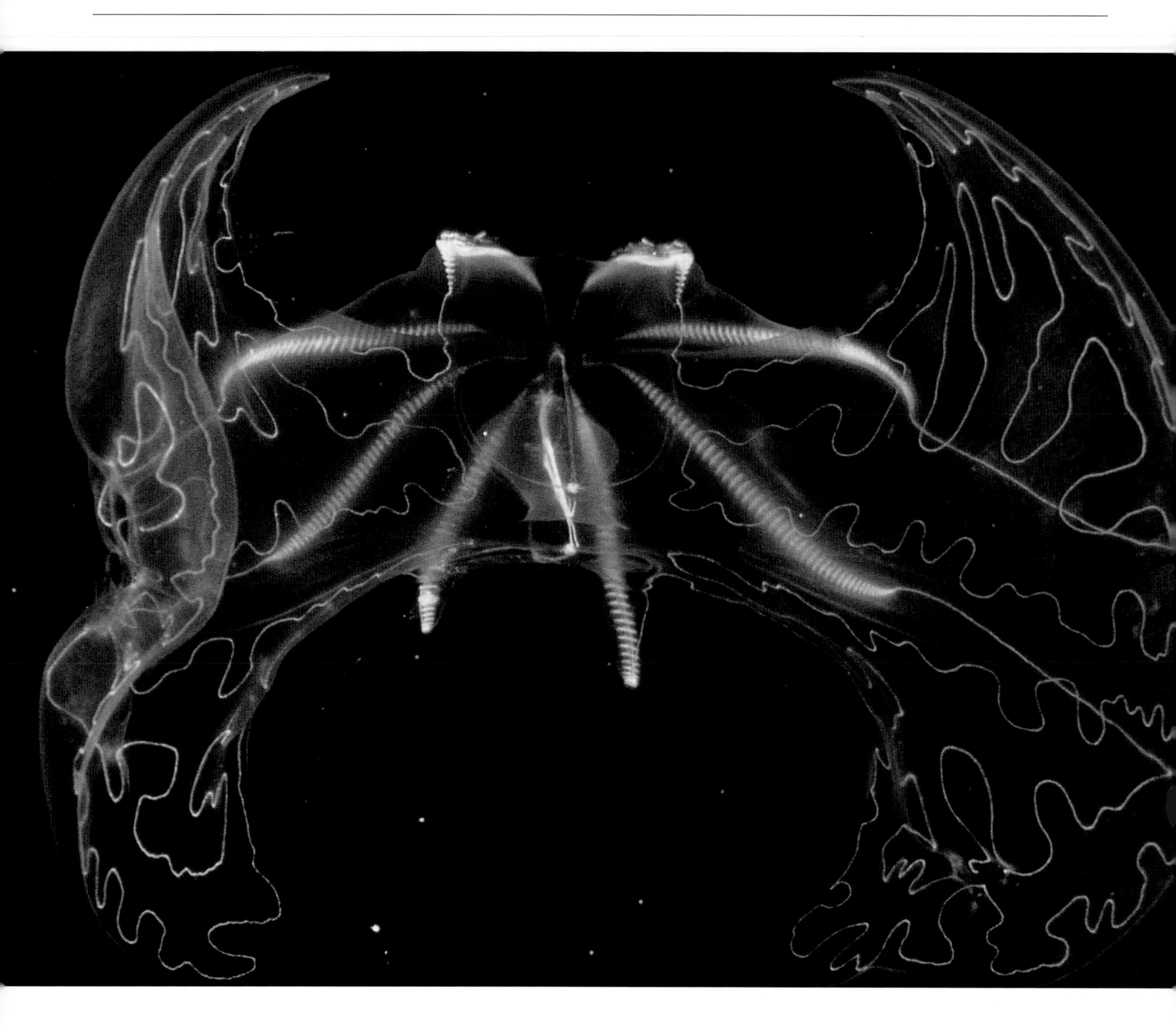

lobes in lyrical movement; some use a mild jet propulsion. The style of locomotion depends on each animal's anatomy. We cannot decide which is the most efficient or elegant technique among the group.

We smile at the movements of midwater molluscs, which seem mildly comical in comparison. Pteropods and heteropods, those midwater relatives

◂ *This lobate ctenophore, as yet undescribed by scientists, swims with lyrical movement of its comb plates and delicate, flapping lobes.*

▴ *A slow-swimming beauty,* Leucothea pulchra, *takes a horizontal path through the water, trailing its secondary tentacles behind it.*

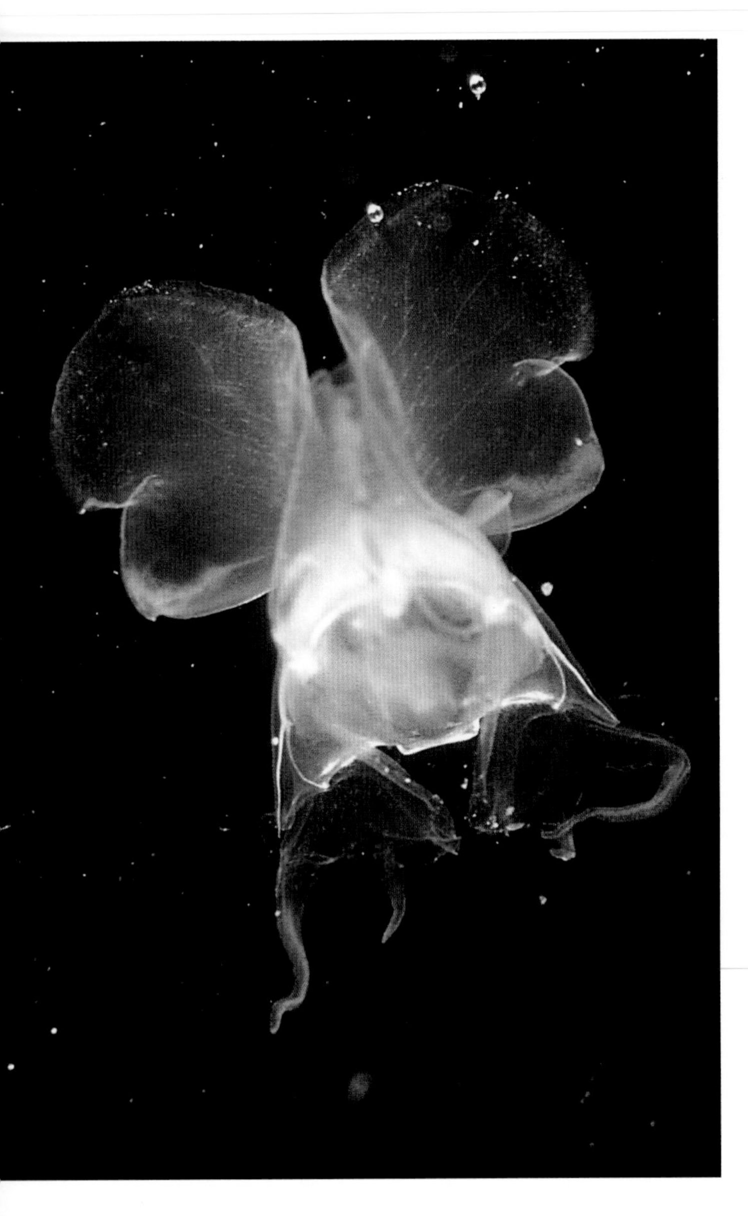

▲ *The winged snail or pteropod,* Cavolinia sp., *makes a mucous web to collect and consume bits of food.*

of snails and slugs, glide through the water like circus creatures. While land snails crawl along on a single foot, the heteropod swims upside-down, flapping its foot overhead. Some, like *Carinaria*, can undulate and flex their body for a quick change of direction. Even when inactive, the swimming snail strikes an artful pose, hanging motionless, head down or curled. Their cousin, the pteropod or "winged foot", swims with distinctive flapping of body lobes that look like wings. We think of them as sea butterflies that flutter in and out of view.

Some jellies migrate vertically to the ocean surface at night and descend back to the depths at dawn. There are freshwater species that make daily horizontal migrations, traveling from one side of a lake to the other. From pole to pole and around the equator, in shallow waters and the ocean depths, we find jellies in nearly every niche. We could extend our search into estuaries, kelp forests and freshwater lakes. We could snorkel over sea grass beds and into mangrove mangles. Not all our quarry would be found adrift—some settle down and attach by suction to rocks or seaweeds. We would find *Cassiopeia* upside-down in shallow tropical lagoons and mangrove canals, pulsing against the muddy seafloor. With stubby mouth-arms pointed toward the sun, they resemble greenish flowers. The symbiotic algae that live inside the upside-down jelly are thus exposed to sunlight, allowing them to photosynthesize and share sustenance with their jelly hosts.

Drifting, pulsing, relaxing, then pulsing again, multitudes of jellies gather in plankton blooms. Some of these blooms signal the first signs of troubled waters. We've seen tiny freshwater jellies bloom in ponds where they were previously unknown. Were the jellies introduced by accident, transferred

carelessly with hatchery fish, in water buckets or on pebbles? Our work as guardians of these gelatinous living jewels is to watch our own impact on the planet. That is what fascinates me so about the mesmerizing world of jellies.

▲ Cassiopeia xamachana *swims upright, but rests upside-down on the seafloor, sunning its leafy mouth parts in shallow tropical lagoons.*

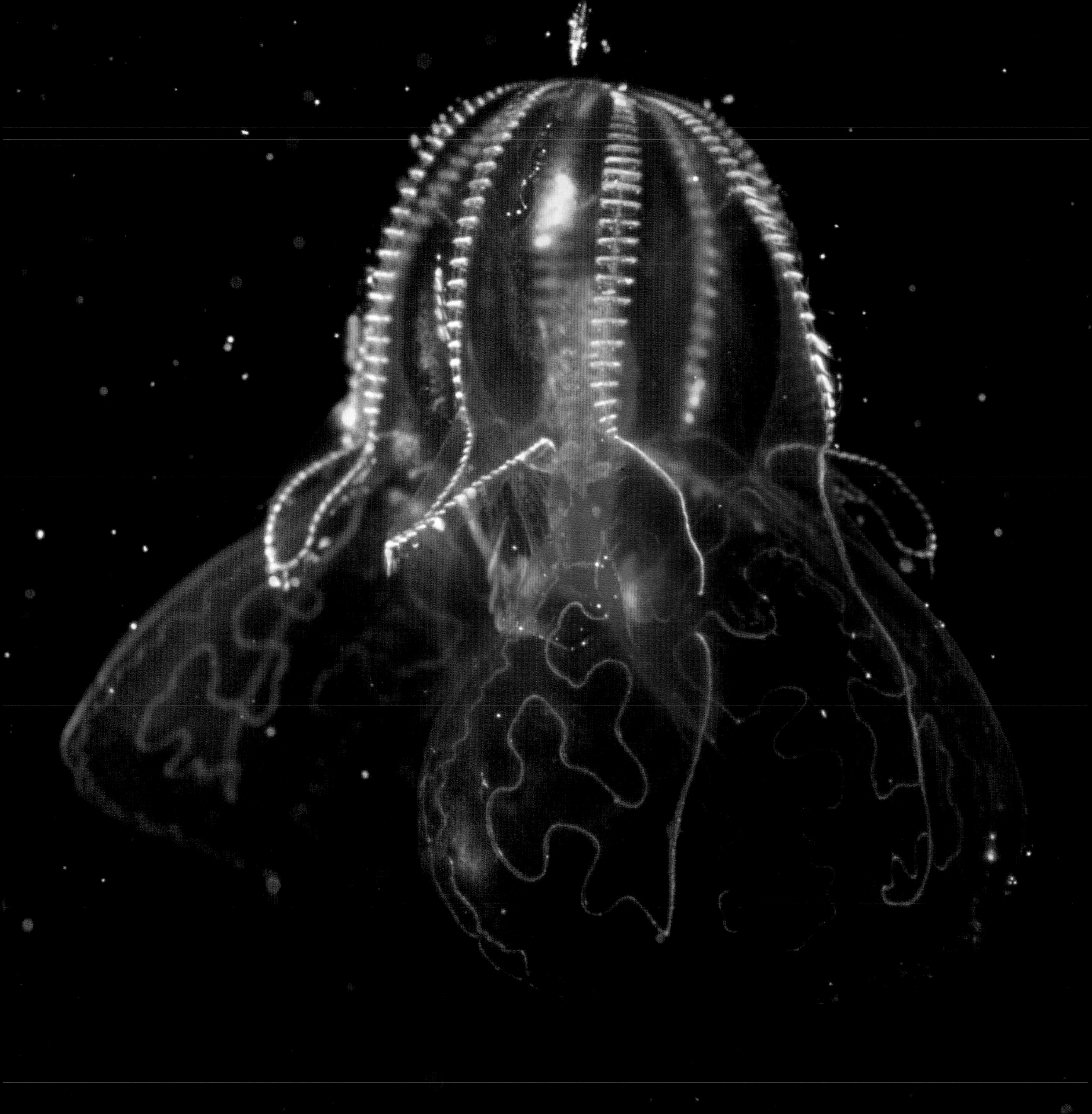

The Language of COLOR AND PATTERN

It is not the language of the painter but the language of nature to which one has to listen.

—Vincent van Gogh

by NORA DEANS

As our guide guns the motor on the skiff, I lean over the side and watch the icy water race against the hull. We fight the fierce current on our way out of the narrow channel. The tide turned while we walked on the island, in the footsteps of grizzlies. Now, the incoming tide challenges our return home. Just as we near the freedom of open water, he cuts the engine. Silently, we drift back, like the ghostly jellies in the clear water below. Their orange color stands out against the dark rocks as they move by. Frilly and full-bodied, they are lion's manes, my favorite. No one speaks. The spell of the moment holds. Again, we start our escape, only to drift back. More jellies. More magic. I want to swim with them, my hair floating in the current like

◂ *The brilliant colors of jellies like* Lampocteis cruentiventer *dazzle in the lights of deepsea submersibles. A "tinkerbell" of a jelly,* Polyorchis haplus, *pales by comparison.*

▴ *A strong swimmer, this demitasse cup size tropical jelly,* Phyllorhiza punctata, *catches zooplankton in small filaments hanging from its oral arms.*

long red tentacles, following the jellies into their world. A watercolor world of mesmerizing colors and patterns.

Everything in Alaska seems painted on a larger and more dramatic canvas, even under the sea. We set out that afternoon with no purpose but to explore the nearby islands. Our companion, an anthropologist turned writer, shares his habitat as an eloquent steward of the northern forests and seas.

I traveled to Sitka in the company of marine scientists and educators. Not a group to spend time indoors, even in the rain, we prowl through tide pools, sail in the company of humpbacks and explore offshore rookeries. And always the jellies are there.

Some are more noticeable than others. Many jellies hide by being nearly invisible. They give the merest hint of an outline encasing nothing but water. Transparent bell jellies are so clear they almost disappear out in the open. Perched on their tentacles in the shallow waters of bays and harbors, they feed on other bottom-dwelling creatures and plankton that drift by. But their clear costumes don't protect them from dredging and trawling and run-off pollution. Bell jellies are fading from heavily populated coastal areas.

Other jellies aren't nearly as shy. Tomato-red with yellow gonads, the jelly *Ptychogastria* also lives on the bottom. In the deeper waters it frequents, red disappears and becomes black in the absence of light. Then why have color at all, I wonder? As I read, I learn that a jelly's color or pattern serves many purposes, and helps it find a meal or avoid becoming one. *Lampocteis* or *Atolla vanhoeffeni* may have opaque, red mouths and stomachs to hide the glow of any bioluminescent creatures they swallow that would, in turn, attract predators.

◂ *Found at depth in the frigid waters of the Arctic and Antarctic and Monterey Bay, the deep reds of* Ptychogastria polaris *disappear in the dark.*

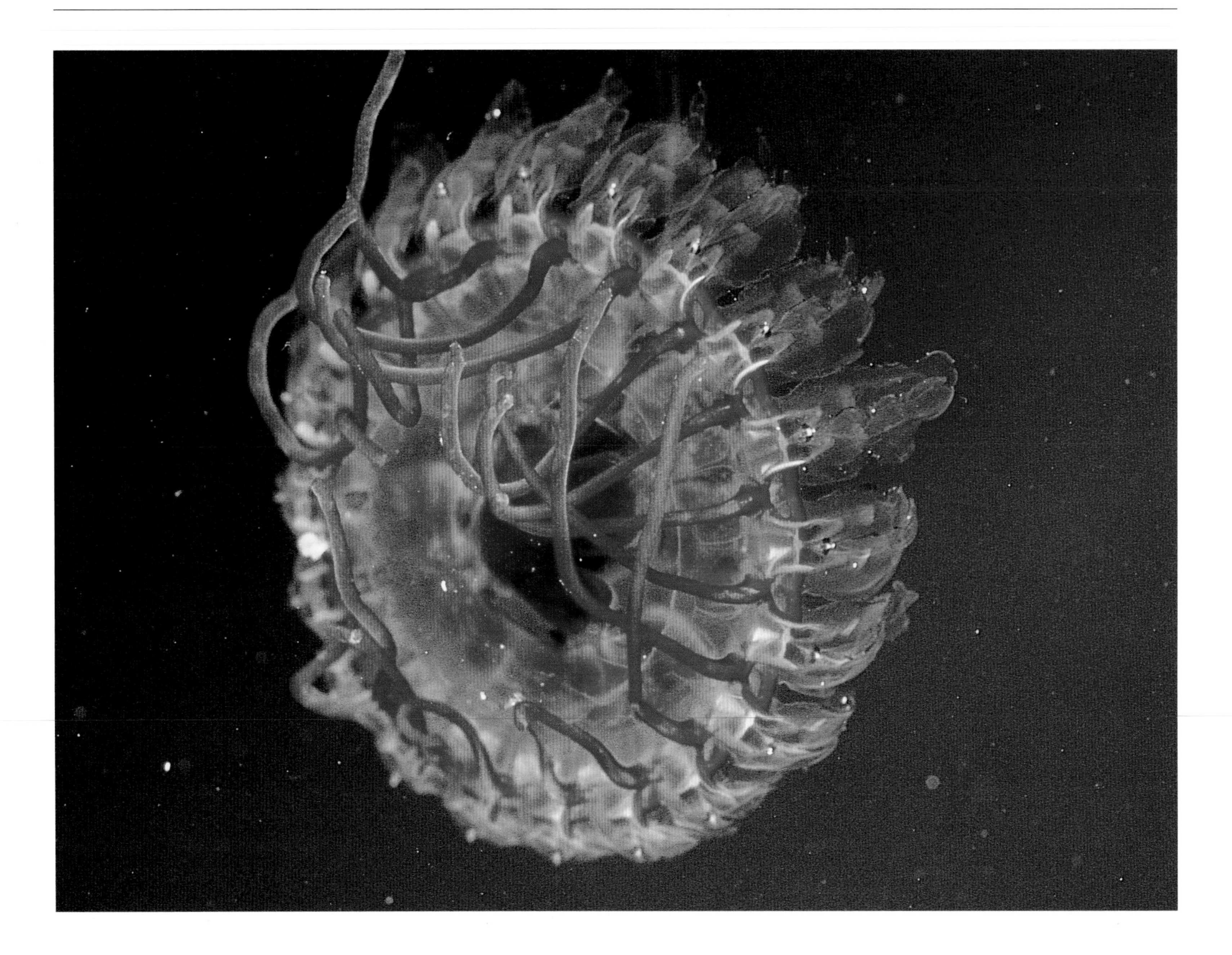

▲ *Close-up, the vivid colors of this* Atolla *take on an appearance of a glass sculpture.*

Every day, we seem to discover some new variety of color or pattern dancing in the sea. Off the coasts of Brazil, Argentina and Japan, colorful flower hat jellies announce their presence with flamboyant green, pink, purple

SEAFORMS, MOCK-UP, 2002 ■ DALE CHIHULY

As sensual as the sea itself, Chihuly's glassforms spark images of marine magic and delight our senses with their beauty, strength and fragility.

Look closely, for within every pattern and swirl you'll find a celebration; this is the magic of colored light revealed.

PHOTO: JAN COOK

CHIHULYANCESTORSEAFORM

PHOTO: MARK MCDONNELL

▲ *Dale Chihuly*

RAPTURE OF THE DEEP ■ RAY TROLL

Plunge into a sensual soup of sea life. With irreverent images, silly humor and real science, Troll's work enlivens the often serious discipline of marine biology.

Look for fish bedecked with different patterns—from spotted and speckled to dappled and checkered. If the colors seem dull at first glance, take another look: you'll find carmine red, vivid chartreuse, burnt orange and royal purple.

PHOTO: CLARK MISHLER

▲ *Ray Troll*

◄ *Colored pencil, crayon and pastel on paper, 3' x 4', 1985.*

and yellow embellishments. It's a good thing you can see them coming, for their sting is nasty and leaves a painful rash. Shrimp fishermen dread them, saying they drive away the shrimp and clog their nets. Small and round with short tentacles, they resemble embroidered and beaded Victorian hats.

Spotted jellies aren't nearly as garish, but their strategy works in the warm Pacific waters they inhabit. Divers love to swim among these harmless

▲ *Moss-green stripes and purple pink bobs set a flower hat jelly,* Olindias formosa, *apart from its kin, and warn swimmers of a nasty sting.*

► *Flotillas of spotted jellies,* Mastigias papua, *swarm near the surface of warm tropical waters by day.*

PHOTO: KATHY MURPHY

PELAGIA, 2001 ■ RICHARD SATAVA

PHOTO: RUDY GISCOMBE

▲ *Rick Satava*

Exquisite in their translucence, these ethereal glass works perfectly mirror the magic of living jellies, creatures Satava encountered for the first time here at the Monterey Bay Aquarium in 1989.

Look for tissue-thin layers of transparent glass and radiant colors glowing within. Fine-spun tentacles seem to drift and twist, as though stirred by the swells of a surging sea.

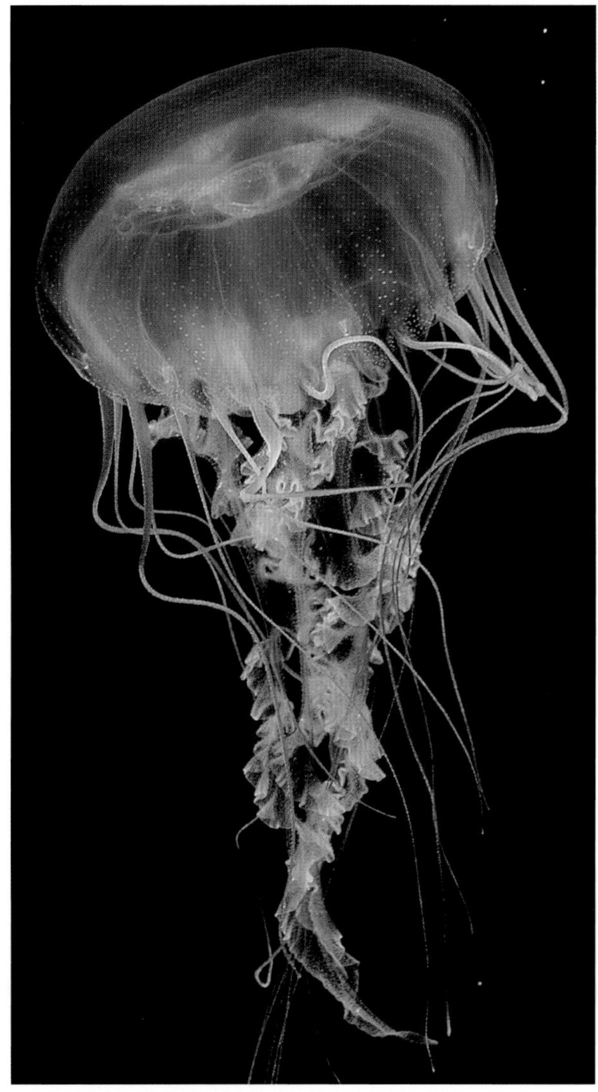

◂ ▴ *Jellies exhibit an artist's palette of colors and patterns, even polka dots, like the ones decorating this spotted jelly,* Phyllorhiza punctata.

▴ *This living, pulsing sea nettle,* Chrysaora fuscescens, *inspires artists like Rick Satava, who translate its exquisite beauty and delicacy into fine art.*

jellies that swarm by the thousands in daily migrations. Compact with many mouth-arms but no tentacles, these jellies farm algae in their tissues, and move constantly to stay in the sunlight. By night, they sink into deeper water where an oxygen-depleted layer, rich in hydrogen sulfide, acts as fertilizer for their algae farms. Some grow quite large, almost a meter in diameter. But spotted jellies are at risk. They live in one of the planet's most threatened habitats—coastal mangrove forests.

Another group of jellies sends out cascades of rainbows as it swims, throwing off so much light it dazzles a predator. One of these, the lobed

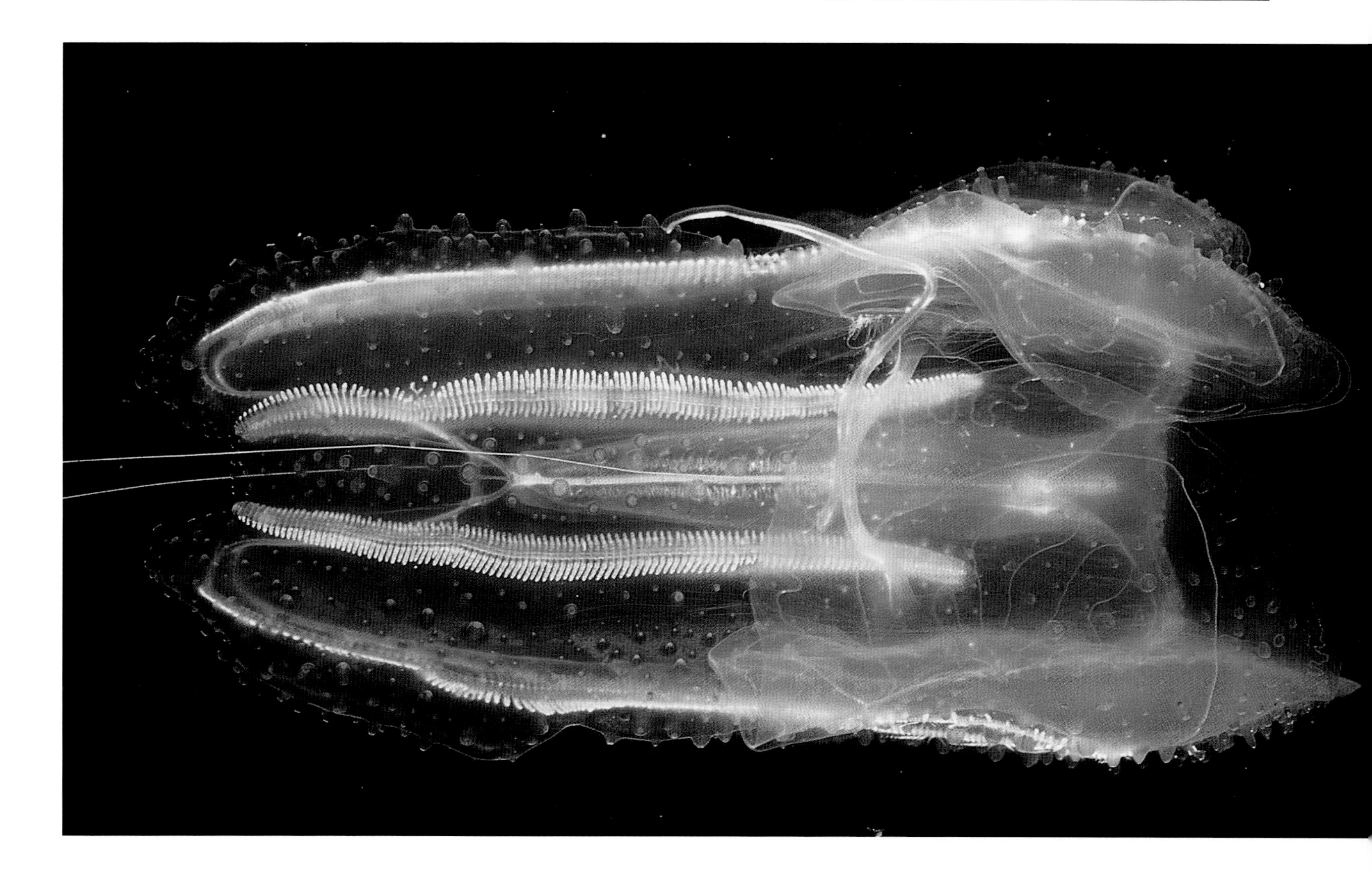

comb jelly, *Leucothea,* is the largest and most delicate in its family. Plankton and other small creatures stick to its large, mucousy oral lobes as it cruises through the water.

Bumps cover the bodies of warty comb jellies, *Mnemiopsis.* Carried from the Atlantic Ocean in ballast water, they were accidentally introduced into the polluted waters of the Black Sea in the 1980s, where they blossomed.

◂ *The bloody belly comb jelly,* Lampocteis cruentiventer, *lives up to its colorful name, putting to shame the paler, delicately lobed comb jelly,* ▴ Leucothea pulchra.

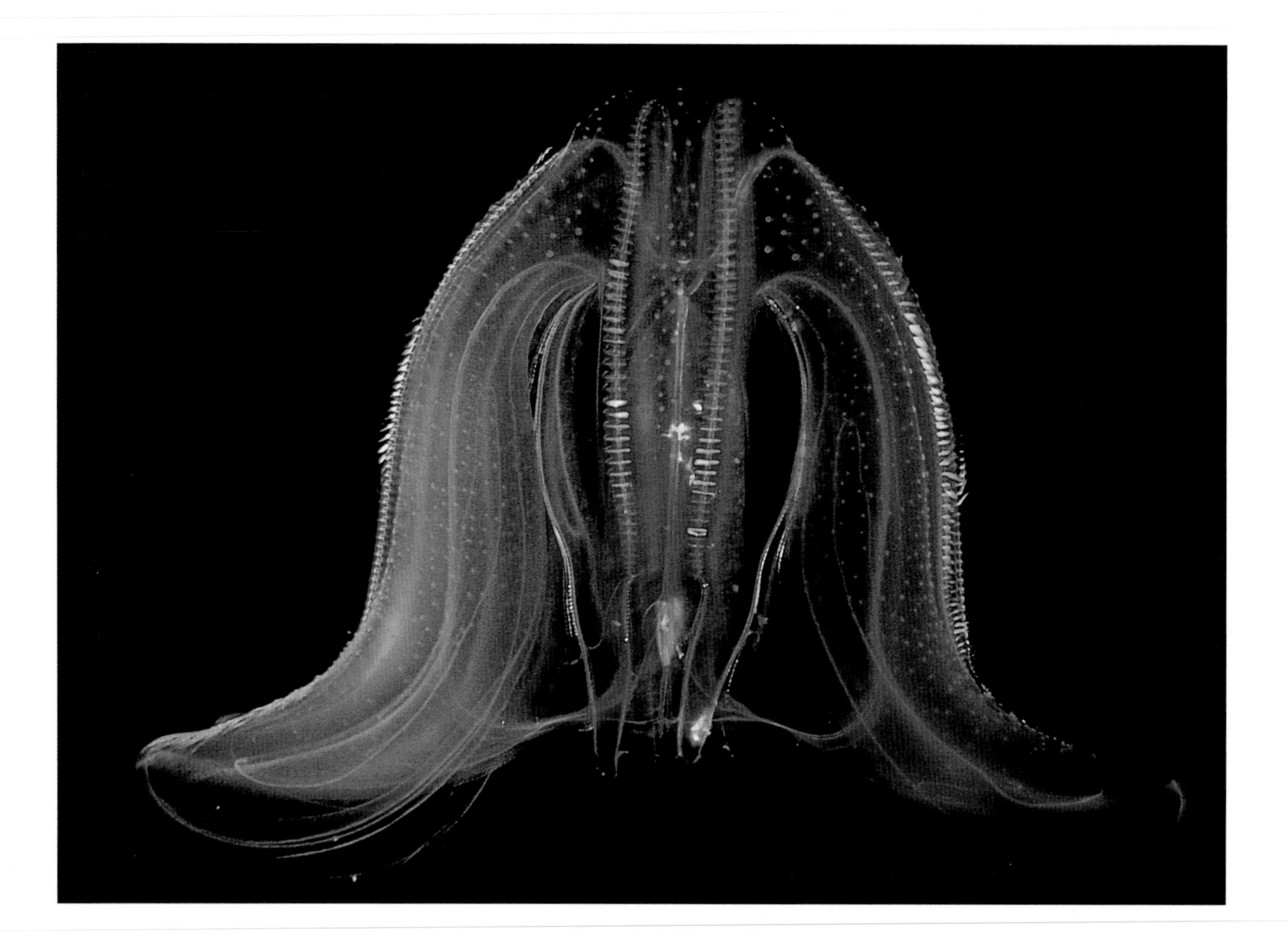

▲ *Voracious warty comb jellies,* Mnemiopsis leidyi, *imported from the East Coast of the Americas now outnumber native jellies in the Black Sea.*

▶ *Sixteen sets of maroon streaks outline stomach pouches of a northern sea nettle,* Chrysaora melanaster.

Billions now infest the sea and devour almost all the plankton. There's little left for anchovies and other fishes. With no natural predators to keep them in check, comb jelly blooms sometimes number more than 3,000 per square meter.

We still have much to learn about jellies and their ocean homes. As Steinbeck quotes Ed Ricketts in the foreword to *Between Pacific Tides:* "'Every new eye applied to the peep hole which looks out at the world may fish in some new beauty and some new pattern, and the world of the human mind

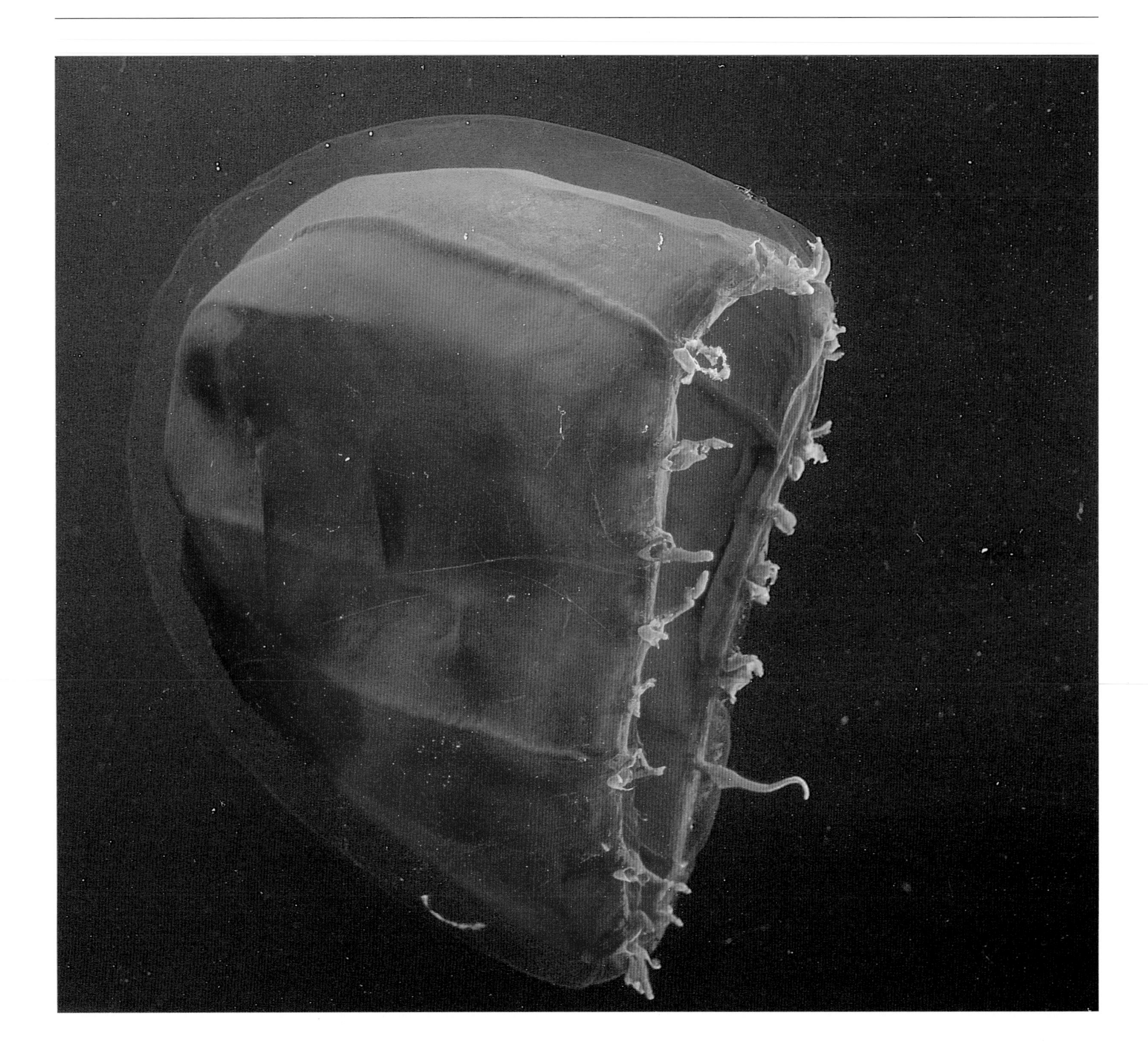

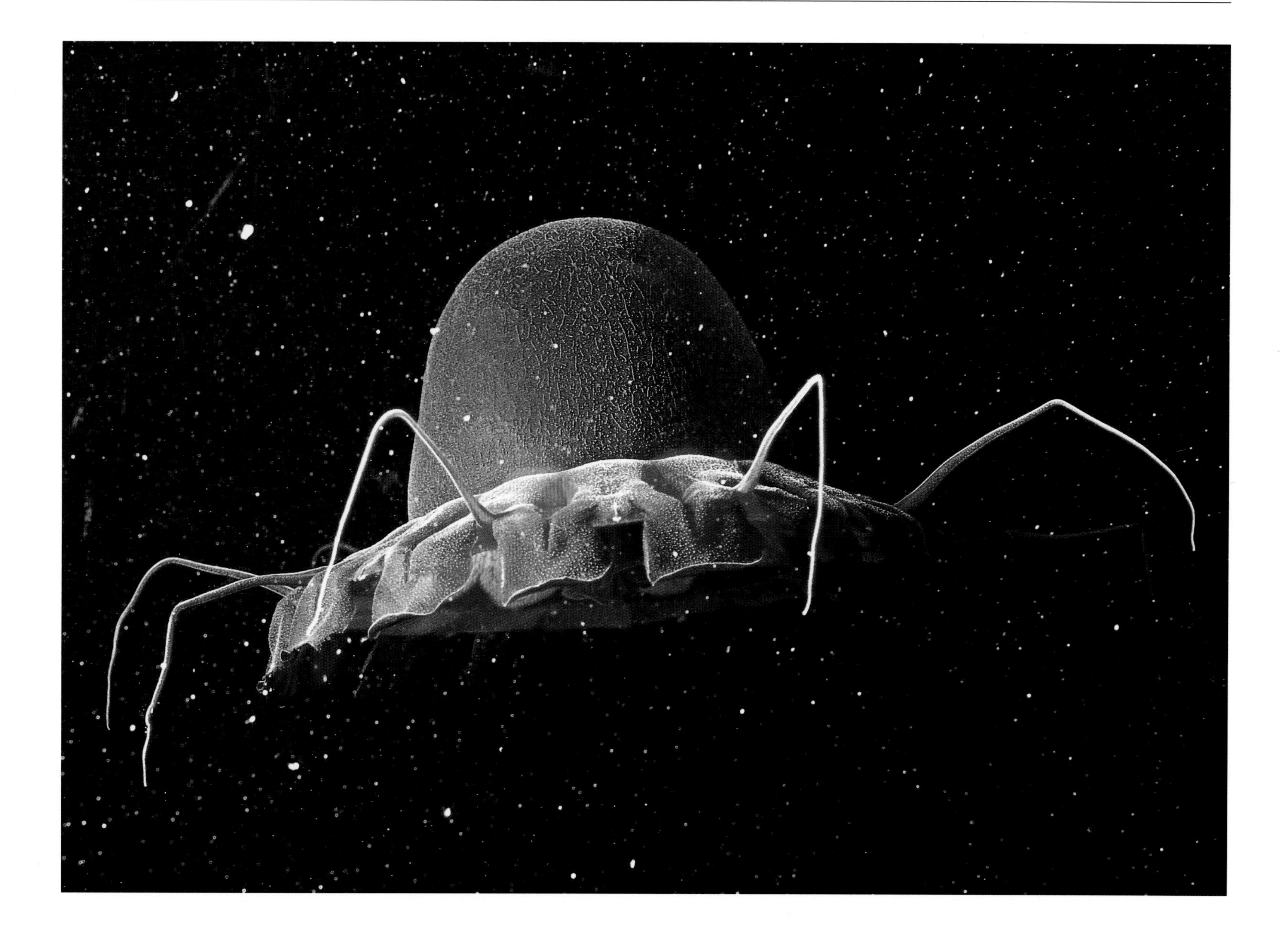

must be enriched by such fishing.'" Steinbeck also wrote, "Look at the animals, this is what we seem to know about them, but the knowledge is not final, and any clear eye and sharp intelligence may see something we have never seen."

◂ ▴ *The dusky deep reds of these midwater jellies make them almost undetectable when the lights are out.*

Secret Stories

INTIMATE OBSERVATIONS

El mar entrega	*The sea offers up*
flores de vidrio	*flowers of glass*
como luz espesa.	*like thick light.*
Son geografias transparentes.	*They are transparent landscapes.*

—Raquel Jodorowski, from *Malaguas*

by JUDITH CONNOR

I am snorkeling between the golden ropes of kelp in Monterey Bay. With a curious novice diver at each elbow, I carefully spread the blades of giant kelp, searching for delicate flower-like hydroids. Clad in thick wetsuits, my two

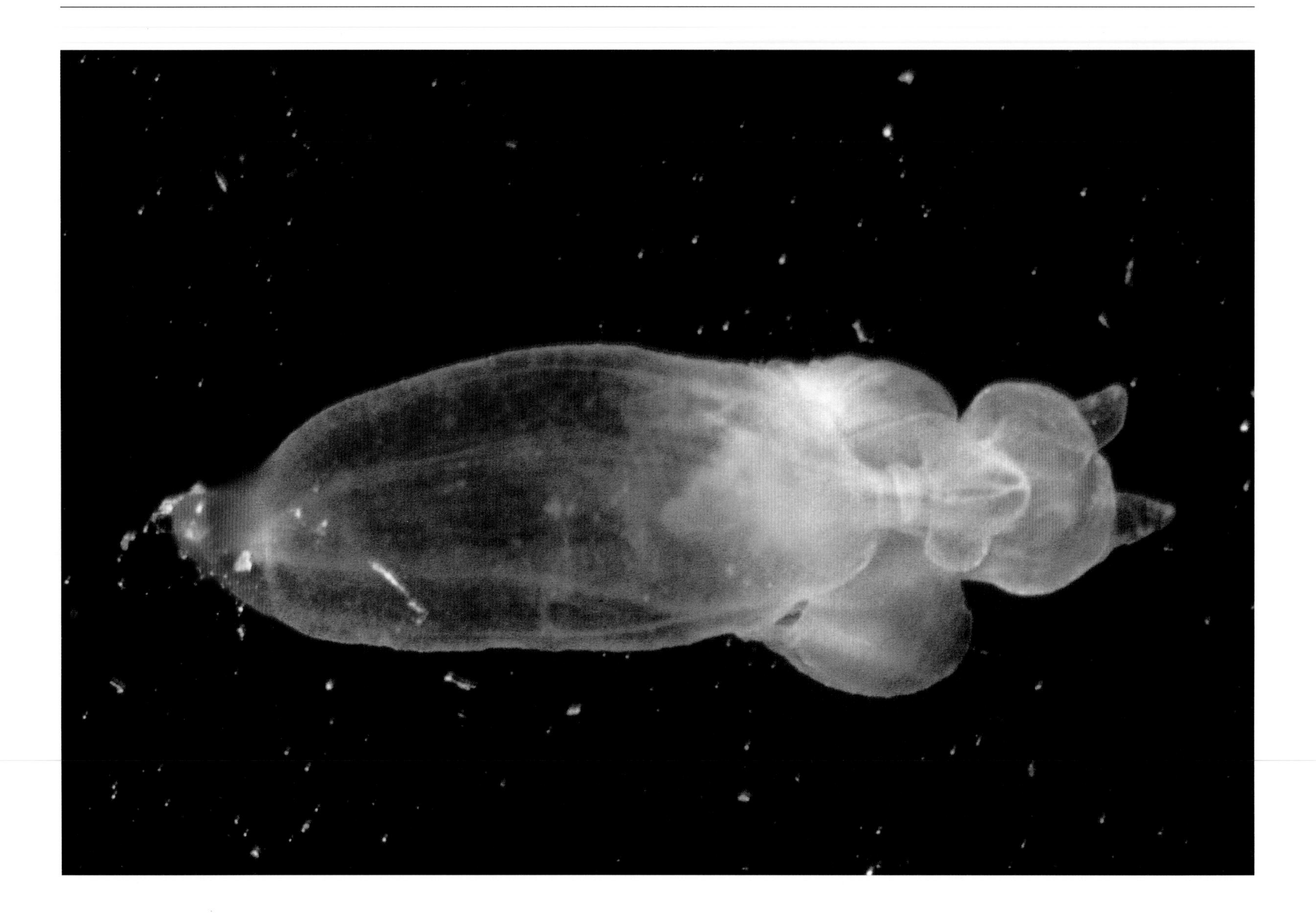

◂ *Bands of muscles that encircle the doliolid allow for its jumpy swimming escape when it needs a quick getaway.*

▴ *Each sea angel,* Clione limacina, *is both male and female, so mating pairs share both eggs and sperm reciprocally.*

students are buoyant and comfortable in the cold bay waters and curious to meet the sea's inhabitants. During spring's upwelling season, this same swim in bitterly cold, green waters would chill us to the bone. In fall, as the winds relax, upwelling pauses and the water warms. We kick out from shore in clear blue and almost balmy seas.

▲ *A tiny larval fish conceals itself between the tentacles of the egg-yolk jelly,* Phacellophora camtschatica.

Tiny hydroids and bryozoans encrust the kelps through many seasons. But fall's warmer oceanic water has swept in unfamiliar species from far offshore: shimmering ribbons, lacy medusae, egg-yolk jellies and graceful sea butterflies. Some jellies, their tentacles nibbled off by blue rockfish, drift aimlessly, stranded near the shore.

▲ *The attached polyps of the sea nettle,* Chrysaora fuscescens, *can bud off to produce small ephyrae that will grow up to become the beautiful pulsing jellies in the surface waters.*

The water is clear enough to see these gelatinous beauties well, but I want a closer look. I take a plastic bag from my wetsuit pocket and snare a few small jellies to take back to our microscopes in the lab.

Dry and warm again on land, I empty my treasures into a glass bowl. We watch the jellies navigate their glass prison; they seem too fragile to survive

▲ *The polyp stage of the moon jelly,* Aurelia aurita, *releases tiny swimming jellies called ephyrae.*

the chaos of the wild. Lights on, my dissecting scope becomes their stage as I peer into the liquid. Delicate golden threads of diatoms remind me of those precious early years of my marine studies. But no labor of love on phyto-plankton today, transparent animals have claimed this hour.

I love the magic of microscopes, the intimate inspection of things too

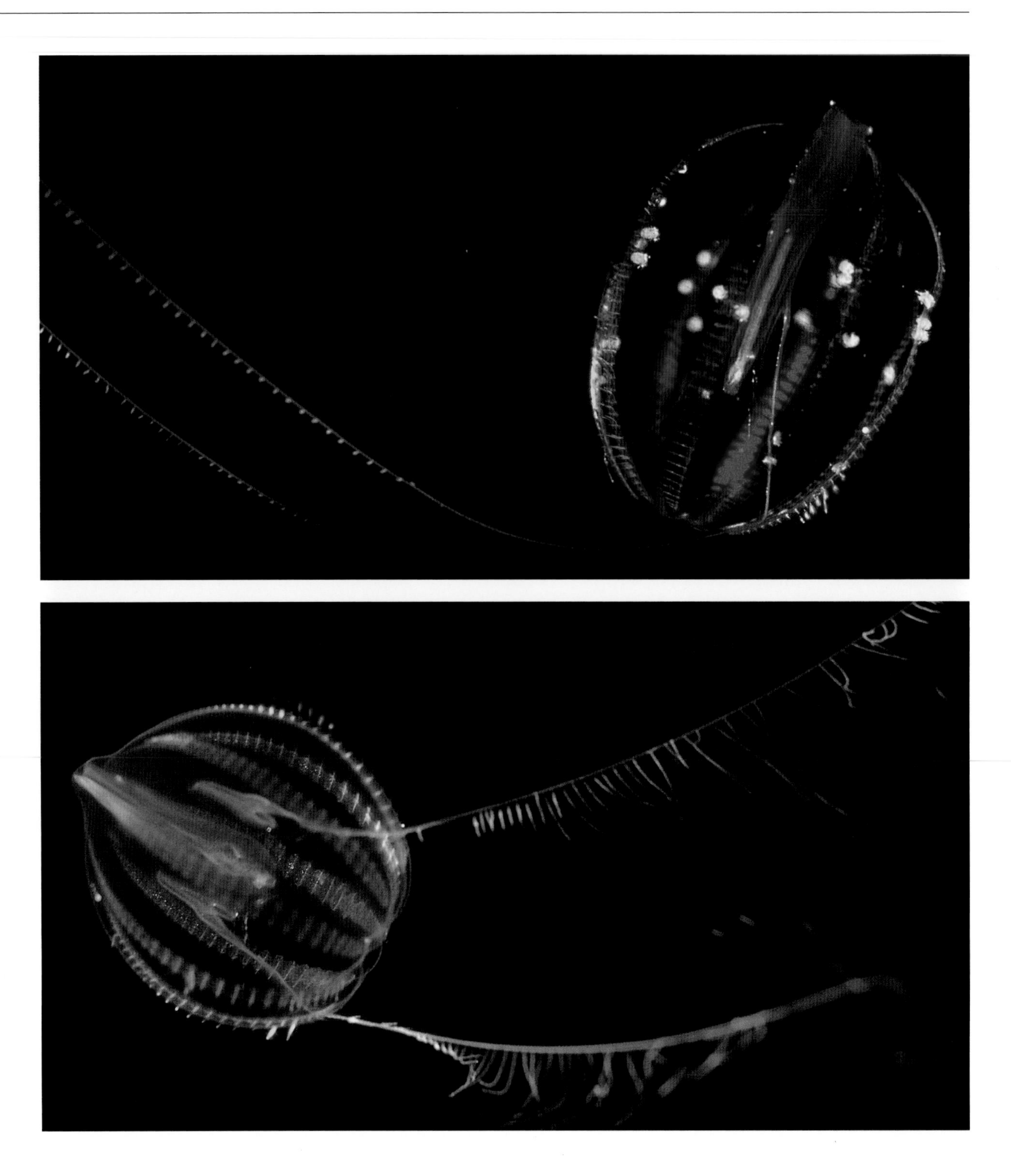

▲ *The sea gooseberry,* Hormiphora, *serves as a drifting home for these tiny, pink hyperiid amphipods.*

▶ *With sticky branched tentacles splayed out like a fishing net, the sea gooseberry,* Pleurobrachia bachei, *waits to ambush small crustacean prey.*

small to see alone. Tiny crustaceans hitchhike on jellies; larval fish hide in their folds. Close-up, the tentacles reveal weapons among the lace. The fringe is armed with tiny capsules called nematocysts. I remember all too well how the box jellies in Namibia left painful welts on my diving buddy that lasted for years. Those nematocysts bear coiled and barbed threads with toxins that can stun or kill prey before they can damage the predator's delicate tissue. I can picture the prey moving from tentacles to mouth-arms, from mouth-arms to the mouth, a graceful, deadly motion.

▲ *The bulge in this transparent comb jelly,* Beroë forskali, *cannot mask the sea gooseberry* (Pleurobrachia bachei) *meal inside its gut.*

Photo: Steven Sloman

Lithographic Water made of lines, crayon and two blue washes without green wash ■ David Hockney

Lines and shadows capture an instant of tranquility in a fluid world that really never stops moving.

▼ *David Hockney*

Photo: Jim McHugh

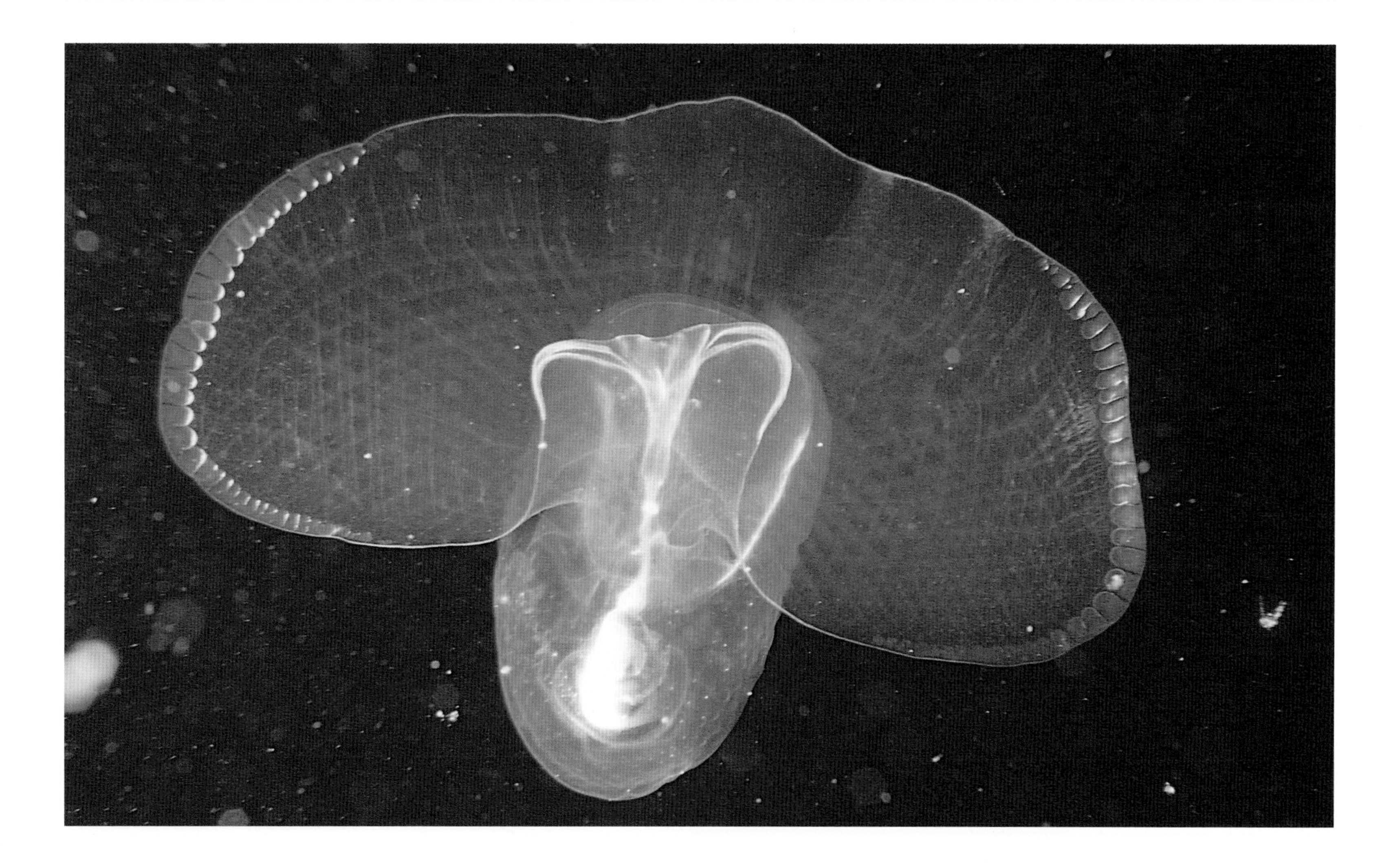

The pulse of the flat jelly creates a current to sweep slow-moving prey into its tentacles. Some tall cones move quickly to overtake their prey. Some are stealthy, lie-in-wait predators. Others lack tentacles altogether and have no central mouth, so they use oral arms to grasp and consume their victims. With my tin of watercolor paints, I try to illustrate their range of feeding styles, but fail to show their pulsing motions. How is it that David Hockney captures watery movement in the simple lines of his art?

▲ *Flapping the winglike extension of its body, the sea butterfly,* Corolla calceola, *slowly ascends through the water, then pauses to drift downward again.*

▸ *Polyps of the egg-yolk jelly,* Phacellophora camtschatica, *strobilate to form stacks of tiny wiggling ephyrae.*

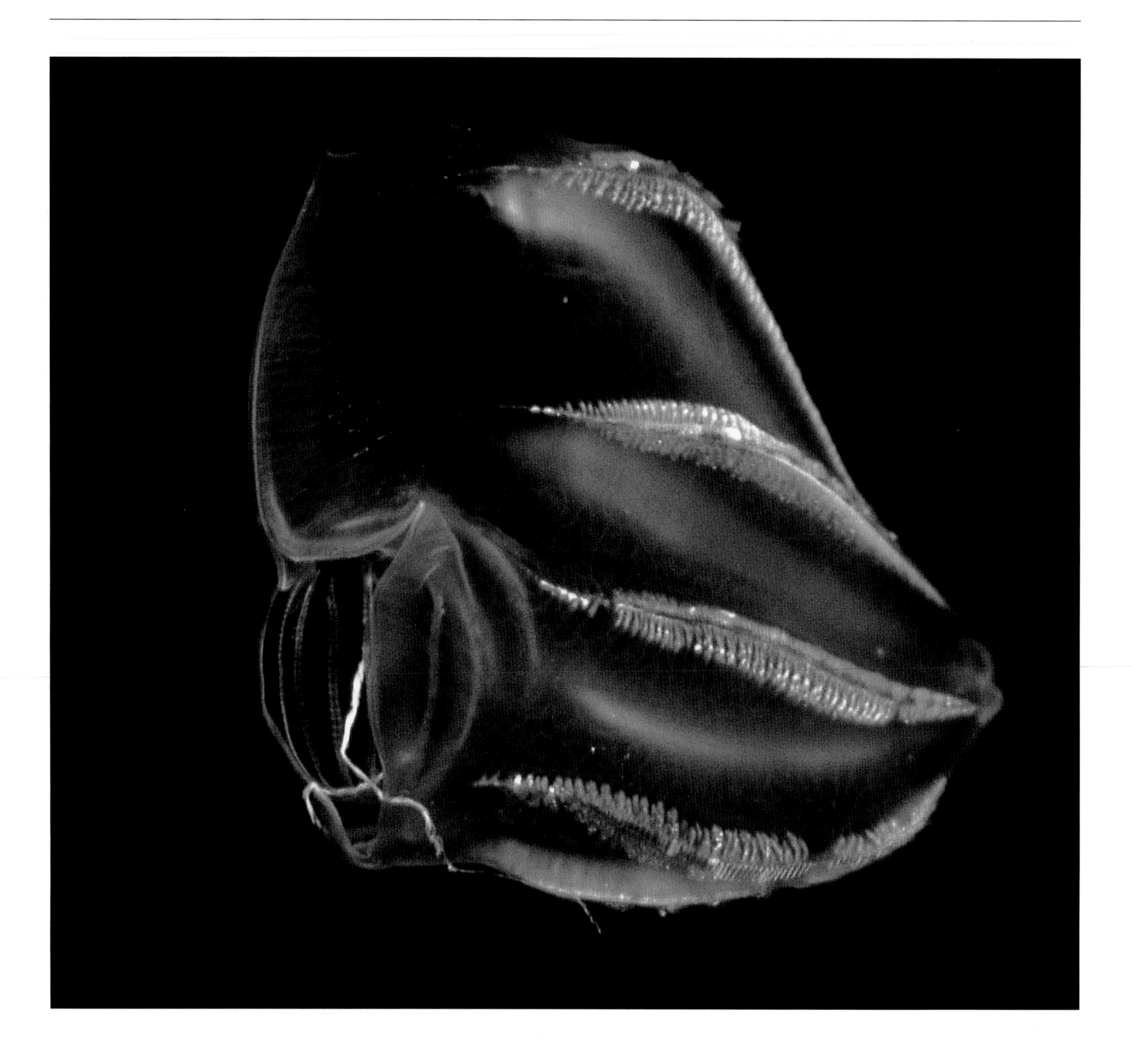

In my mind, I see a sea gooseberry dangle its two fishing tentacles, whose sticky tips snag prey. This comb jelly finds its foe in *Beroë,* a comb jelly without tentacles. Tiny triads of teeth inside *Beroë's* mouth are razor-sharp and ready to tear into gelatinous prey.

I think of Rachel Carson's description: "Late that night came the legions of *Beroë,* the cannibal ctenophore, a sac of pinkish jelly large as a man's fist. The tribe of *Beroë* was moving out into the coastal waters. . . . The sea brought them to the place where hordes of *Pleurobrachia* lay twirling and quivering. The big ctenophores fell upon the small ones; they ate them by hundreds and thousands."

Nature proves the profit of eating one's watery kin. *Aequorea,* for one, eats other medusae—bioluminescent ones that use their own enzymes to make light. The prey provides the crystal jelly with a promise of its own light-making.

One of the jellies in my vessel carries packets of small transparent eggs. Given the chance, this female will cast her eggs into the sea to mix with sperm from a male of the same species. The mystery is how these sightless mates find each other. What becomes of the drifting eggs? I look over to another vessel in my lab that holds the clue—we can grow the eggs in culture. The eggs grow into polyps that spread across the glass as well as they would on the seafloor—or on floating seaweed or shells. The final stage to this secret story is when the captive polyps develop chains of new bells. These tiniest bells bud off in sequence, a process we call strobilation. I roll the word around my tongue, *strobilia*—Greek for overlapping pinecone bracts. A word can hardly capture the amazing sight—a writhing stack of jellies that wait in turn to be set free.

◂ *A doomed* Pleurobrachia bachei *is just visible inside the open mouth of the predatory comb jelly,* Beroë forskali.

Inspiration

NATURE AND ART

Can a painting be a prayer?

Can wilderness be a prayer?

—Terry Tempest Williams, from *Leap*

by JUDITH CONNOR

I find my restoration in nature. Some of my sensual pleasures may be peculiar to me alone. Perhaps no one else would value the same aspects that I do. When I hike the foothills of the Sierra Nevada, I go purposely to finger lichens on granite boulders. Hauled up on the sand after an ocean dive, I relish the pungent, acid smell of beach wrack. There's simple pleasure in the sunlight on the slough, its warmth on my face a contrast to cold dampness on my legs. My kayak, rafted with those of my writer friends, drifts as we feast in our precarious positions. We pass olives and cheese across our kayak skirts, pointing to sea gull chicks on the shore, and golden sea nettles that float beneath us.

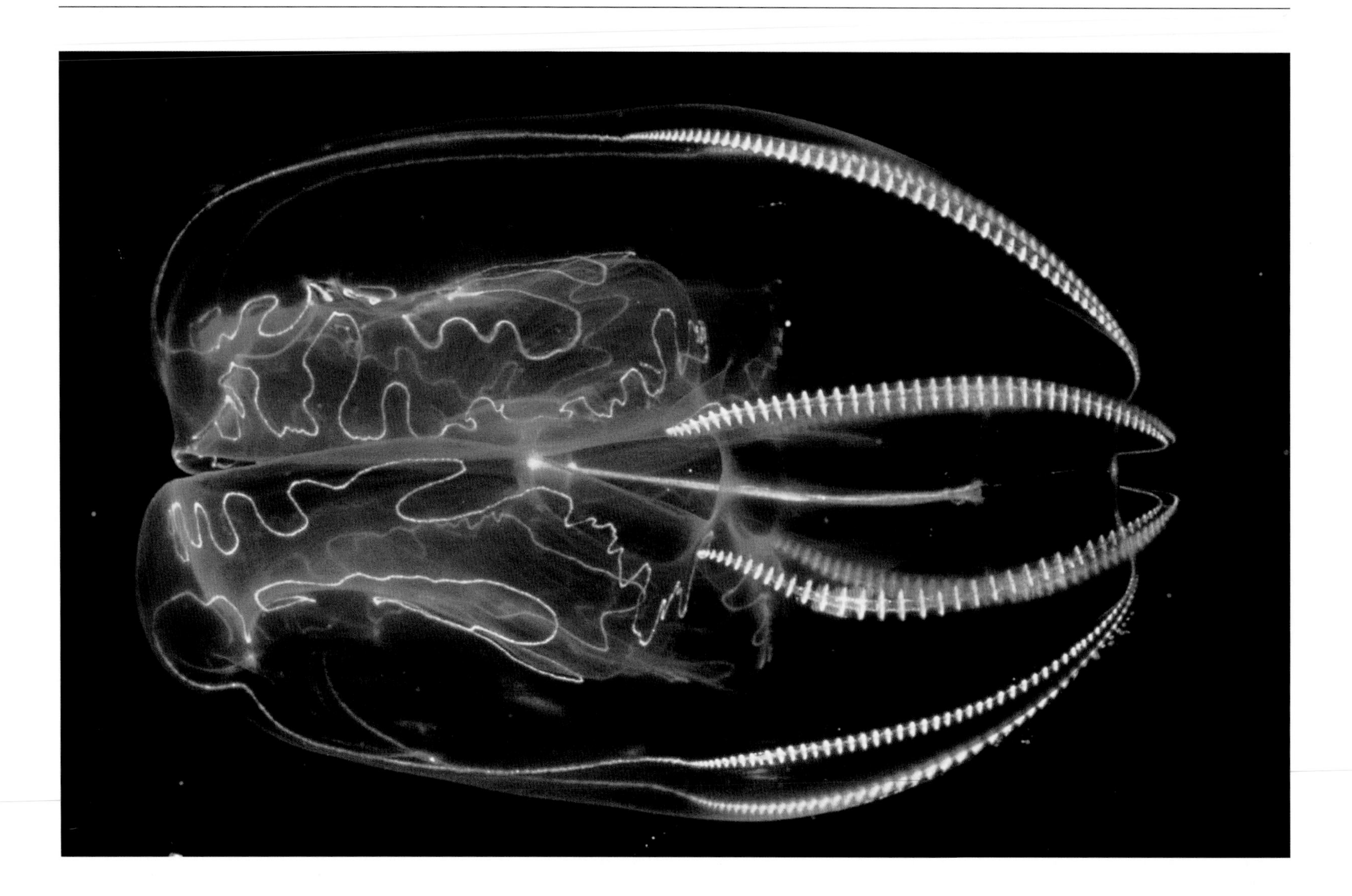

◂ *Algae living inside* Mastigias papua *provide additional nutrition to the jelly enhancing its plankton diet.*

▴ *Because delicate lobate comb jellies are easily damaged during collection, some species have not yet been described by scientists.*

Lucius Seneca held that "all art is but imitation of nature," but I'm not sure that's always true. Adrift, we eat and talk of experiences and art that have inspired us. Art restores my spirit, but usually in ways unlike nature's sensuality. I search out images that startle or make me smile; I find strange humor in Roger Brown's paintings of unexpected scenes, fish scales and silhouettes, drive-ins and diners. It interrupts the predictable.

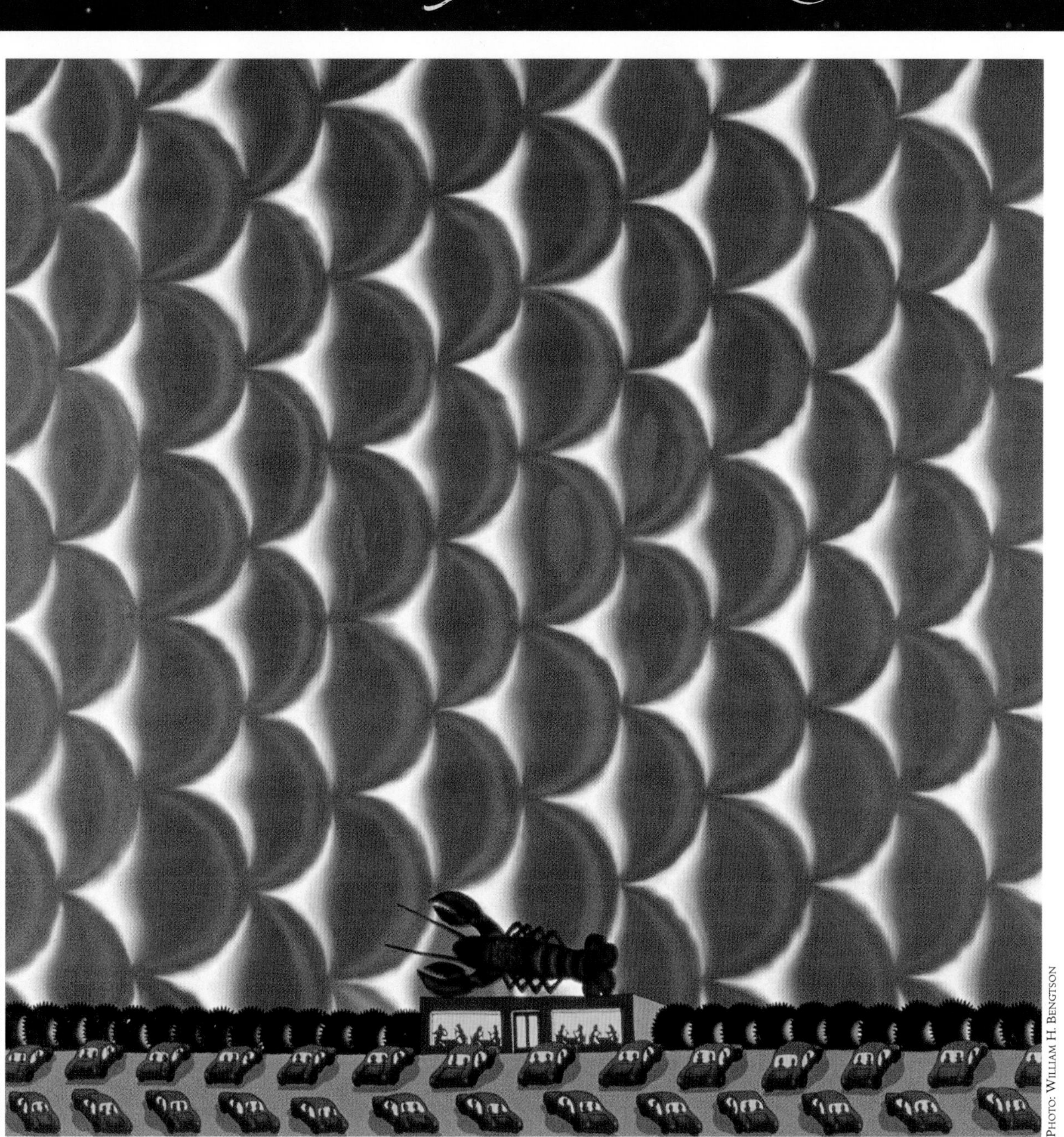

COURTESY OF THE SCHOOL OF THE ART INSTITUTE OF CHICAGO AND THE BROWN FAMILY

PHOTO: WILLIAM H. BENGTSON

WHERE HAVE ALL THE FISHES GONE? 1991 ■ ROGER BROWN

Shadowy figures outlined in windows, grimly lit restaurants and an atmosphere of mystery—you're seeing typical features of Brown's colorful paintings. Can you see how more than three-quarters of this painting portrays a pattern of shimmering fish scales?

◀ *Oil on canvas 72" x 72"*

▼ *Roger Brown with his painting,* Lake Effect.

©1992 JIM ARKATOV

▲ *A polyp of egg-yolk jelly,* Phacellophora camtschatica, *strobilates tiny, writhing jellies to be released in sequence.*

▲ *Scarlet lips mark its focal point, the stomach of the deep-sea jelly,* Atolla vanhoeffeni.

I love art that makes me ponder, art that makes me rethink my role in life, my responsibility to do something meaningful. My favorites have natural elements in their designs. Andy Goldworthy's constructions of rocks, sticks and snow are of nature and often in nature, but unnaturally positioned. The hand of the artist is there in a snowball he gathered below an ash tree, took into his studio, stained with dye extracted from ash seeds, and melted on paper. What genius in his combination of media that yields sliding wet neutral contours on that paper! I'm reminded of a gelatinous sea creature, once lively, now dying, drying, disintegrating into transparent tissue.

In Lanny Bergner's sculptures I detect biology disguised, enhanced with hardware. Is there a vegetable gourd under that wire, twisted aluminum screen and glass? I'm drawn to his strange wraps; I recognize that one is made

COURTESY OF ELLIOTT BROWN GALLERY

PHOTO: CLAIRE GAROUTTE

FUNGOIDS I, III, II 2000 ■ LANNY BERGNER

Like strange animals and plants still growing and evolving, Bergner's sculptures echo the infinite variety of organic forms in nature.

▼ *Lanny Bergner*

PHOTO: EVE DEISHER

▲ *As the comb jelly,* Beroë forskali, *swims, rows of cilia move in synchronized waves and diffract light into a rainbow of colors.*

► Bathocyroe fosteri *spreads open its oral lobes to capture food on the sticky surfaces that surround its central mouth.*

from kelp, the next seems to be plastic. Walking through a Bergner installation, I think he must have seen real deep sea creatures. I recognize the shapes of rare siphonophores, the comb rows of ctenophores in his designs. I'm drawn to defend nature's superior way with beauty.

Nora queries and reminds me of glorious tinted glass, that human invention. "Remember how Dale Chihuly captured the transparency and color of medusae and froze their undulations in glass?" she asks. Has he seen the sapphire comb jelly? Does he know the throbbing tomato-heart of a jelly we call *Atolla?*

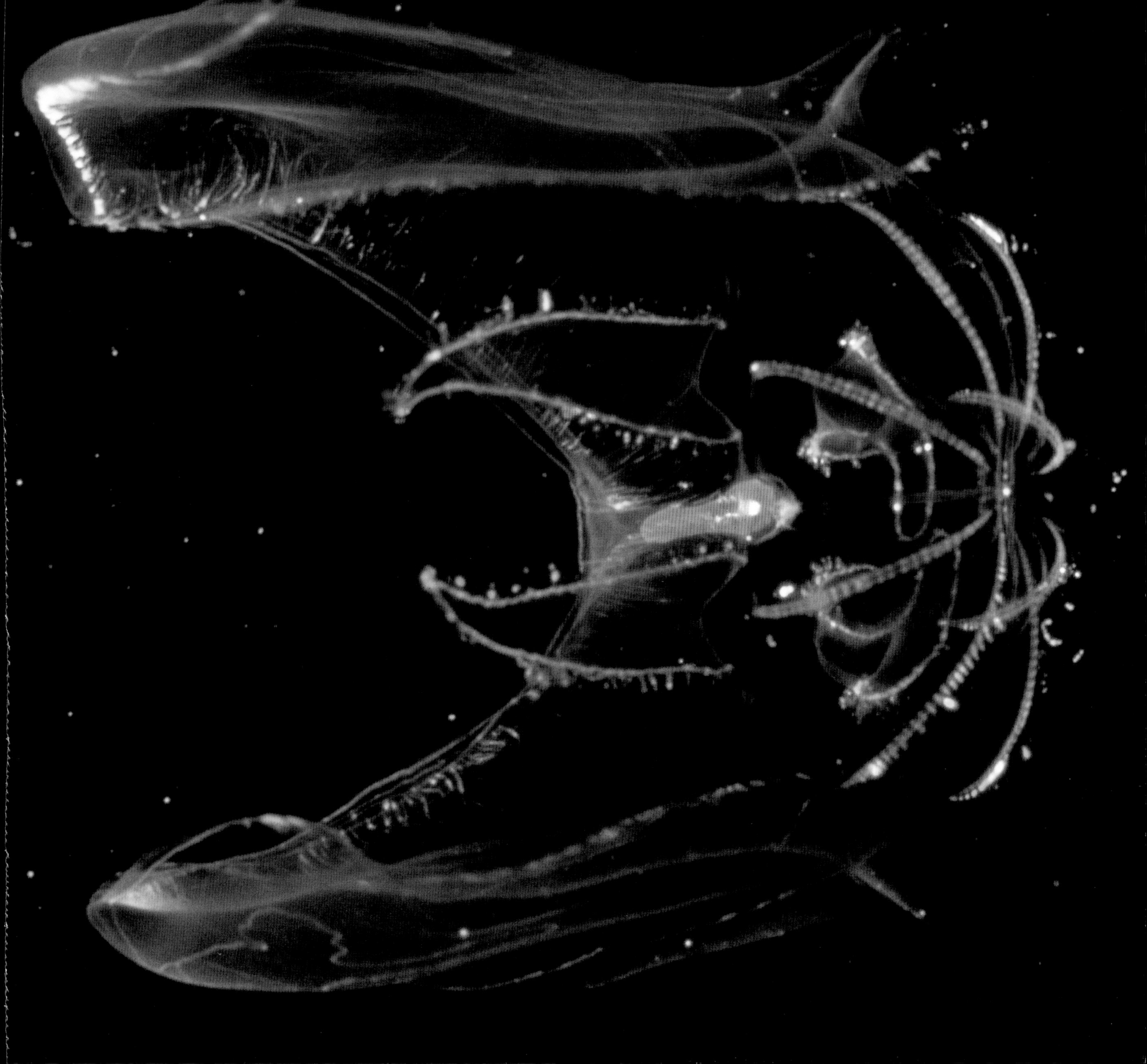

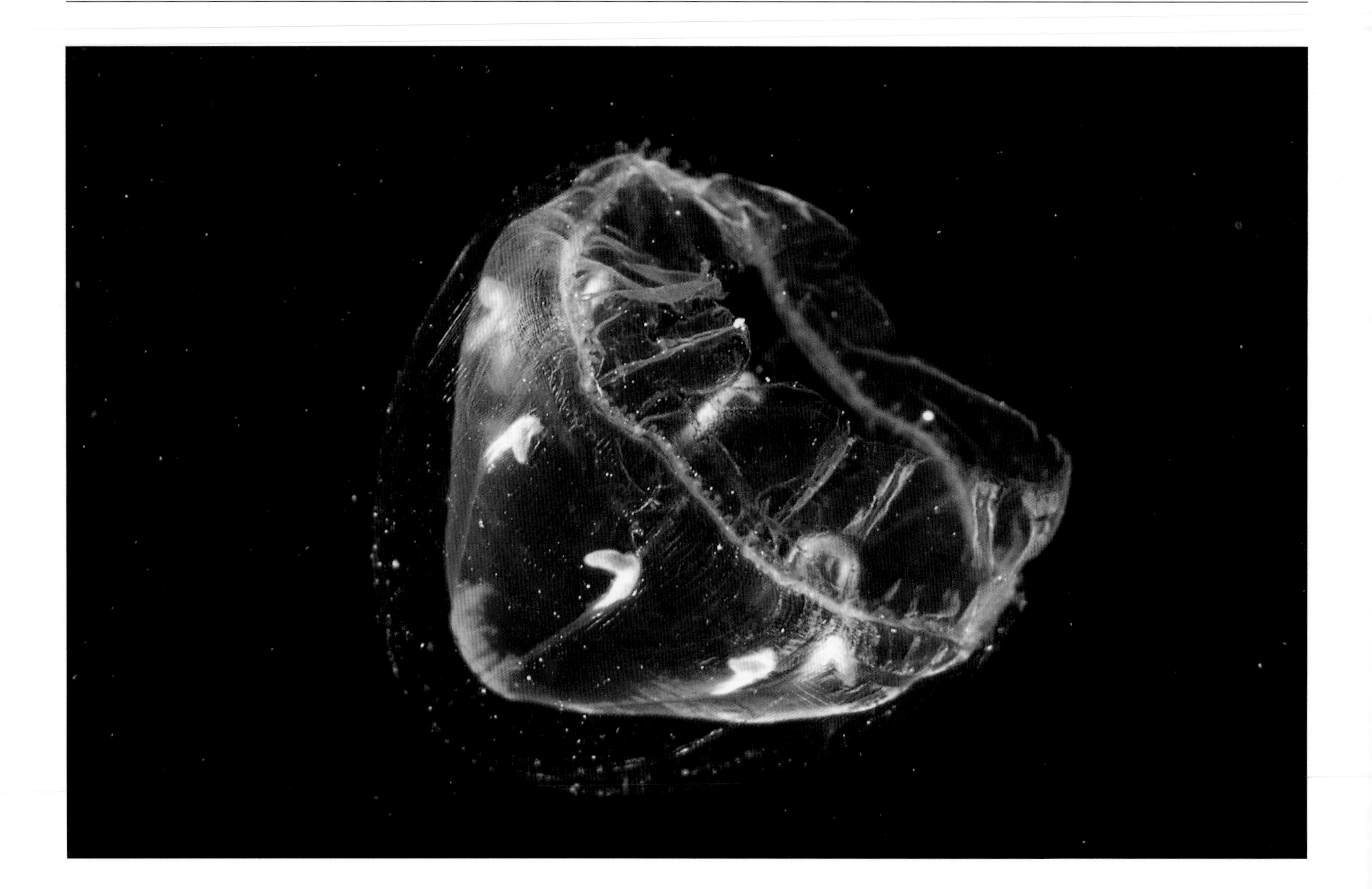

▲ *Deep sea biologists have discovered that the beautiful* Crossota alba *is bioluminescent and makes its own light.*

Marc Chagall said that "great art picks up where nature ends." I don't want to separate the two—I want both. Invite Cork Marcheschi to launch his glass discs overhead with lights that project patterns on the walls around me. Although Marcheschi's art evokes the bioluminescent jellies, his colored lights show art from chemistry and physics. The names of inert and noble gases that he employs sound like poetry to me: neon, xenon, argon and krypton.

Photo: Anna Bradley

▲ *Cork Marcheschi*

Photo: Aengus McGiffin

These glass sculptures respond to a touch with wave upon wave of flickering light. Gases—like neon, argon, xenon and krypton—surge within the sculptures to create dancing lights.

Liquid Luminous Secrets, 2002 ■ Illuminated glass pieces created by Cork Marcheschi, Jim Nowak, and Reid Johnston

▲ *Matt Gray, self-portrait*

Gray's photographs depict jewel-like portraits of the everyday, ordinary lollipop; they seem to give off an internal glow, a tasty vision.

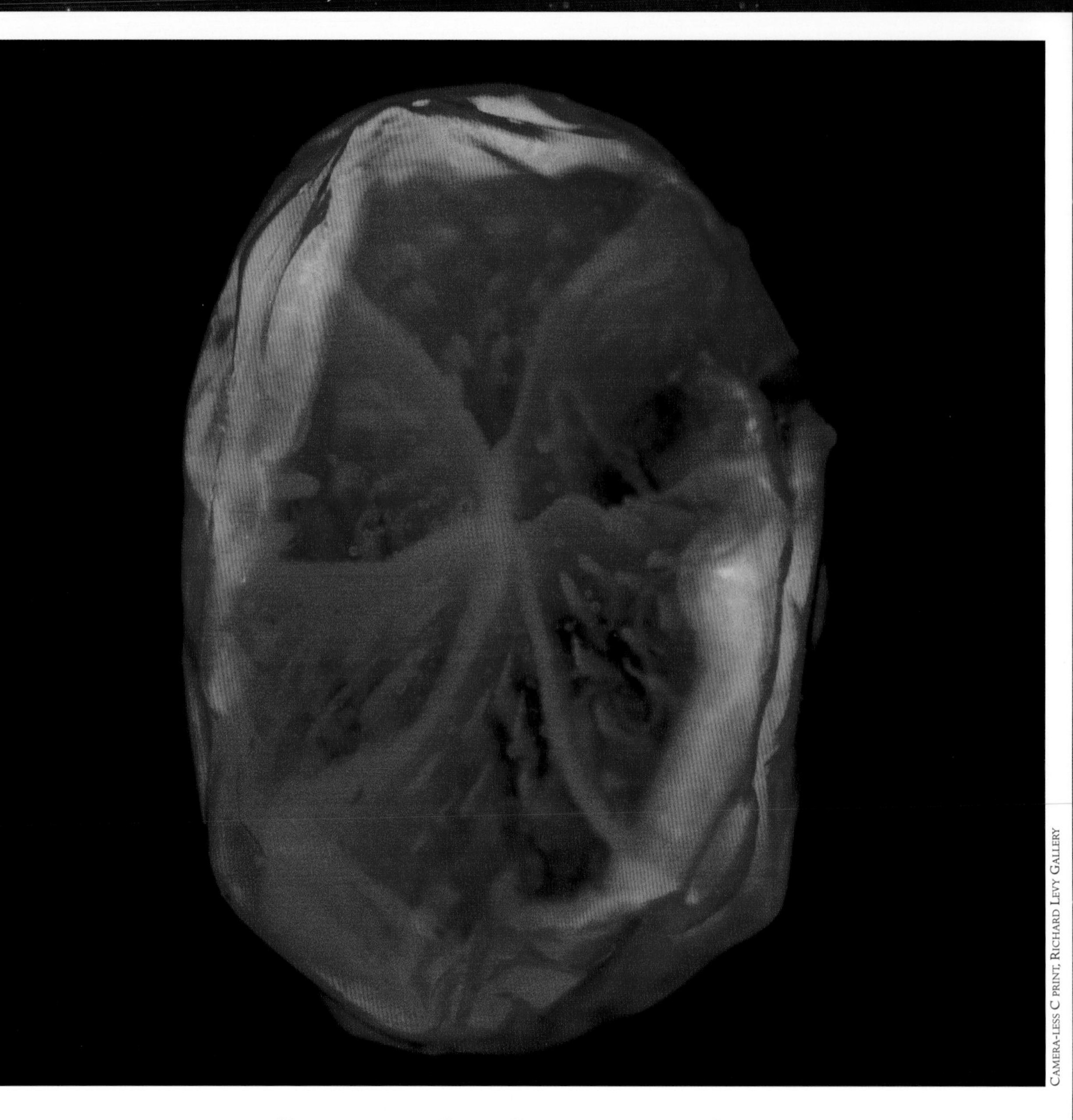

CAMERA-LESS C PRINT, RICHARD LEVY GALLERY

UNTITLED – FROM STUPID CANDY, 2000 ▪ MATT GRAY

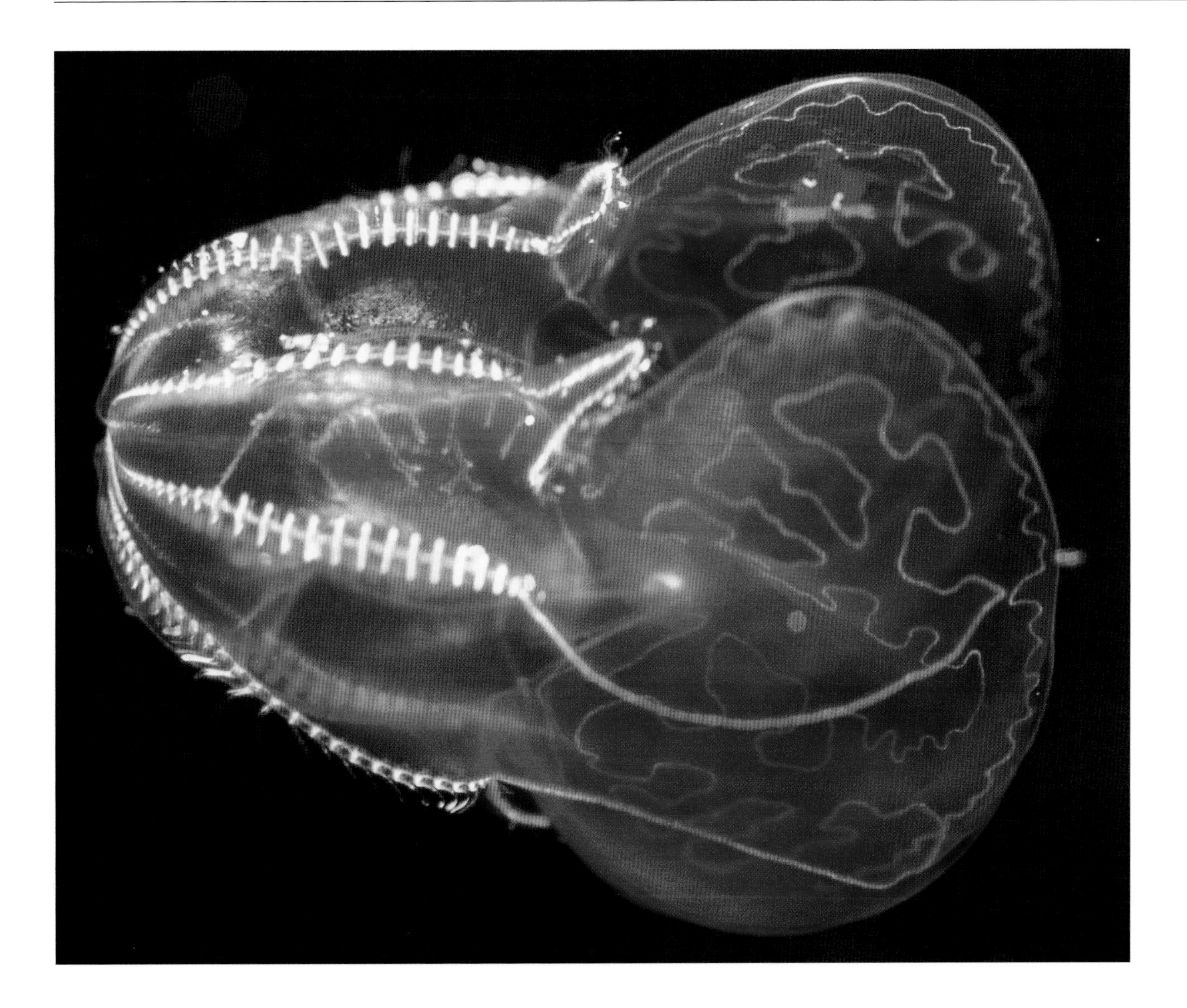

▲ *This lobate ctenophore comes in various hues, ranging from clear to red to a deep purple, but the gut is always blood red, hence its name "bloody belly".*

Suspend Lanny Bergner's sculptures above my head. Surround me with walls of Matt Gray's luscious, jewel-like lollipops. And show me the living, swimming creatures on display, too. Preserve a place and time for contemplation. Reflect on our roles and our role models in science, art and conservation: Julie Packard launching an exhibit of living and fine art;

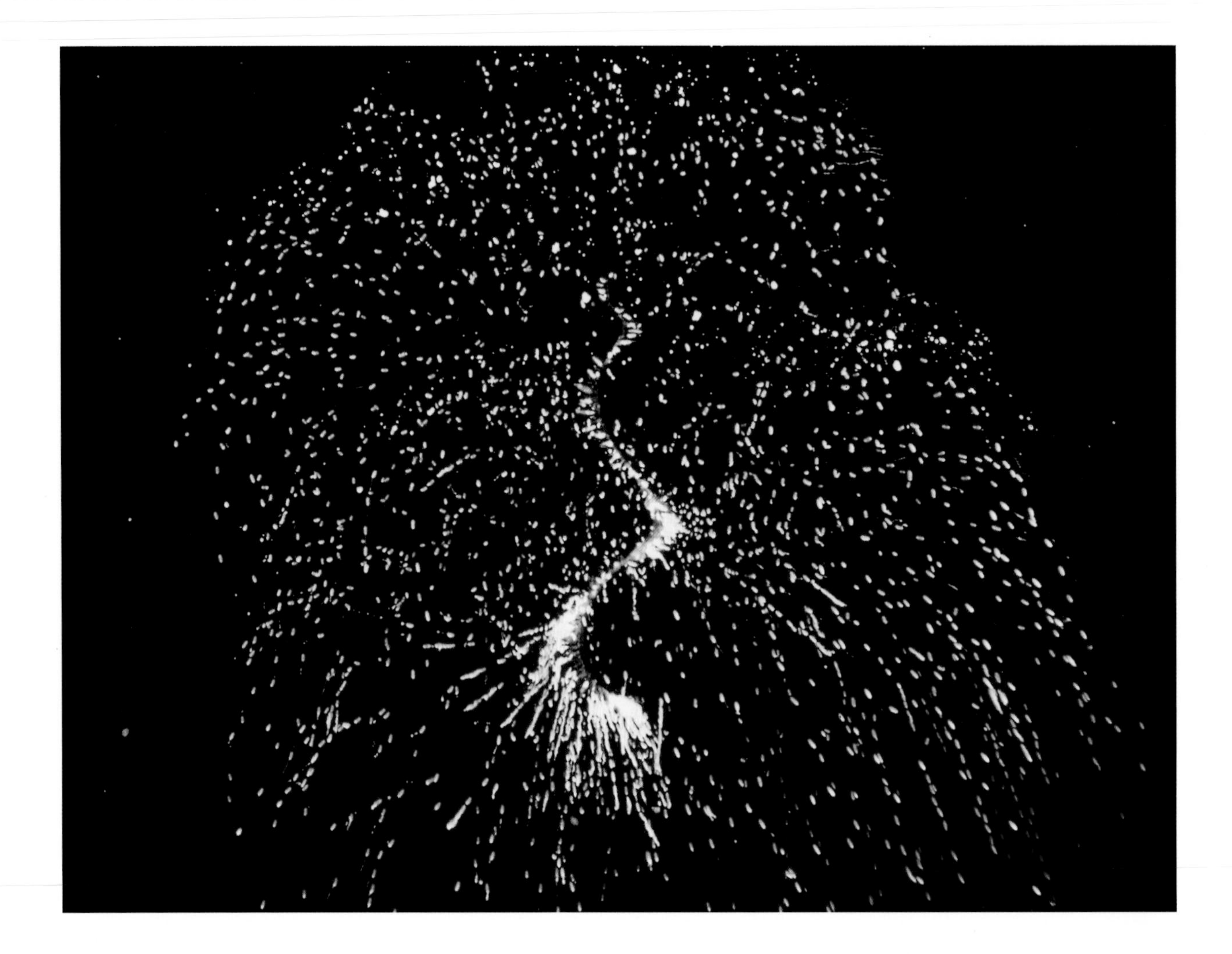

▲ *An undescribed siphonophore lights up the deep sea like fireworks in the night sky.*

▶ *With long tentacles adrift around its sticky oral arms, the egg yolk jelly,* Phacellophora camtschatica, *captures and eats other jellies.*

Marcia McNutt reviewing research video of pulsing salps; Nora in the kayak next to me sharing olives and opinions. Remember the open ocean, owned by no one and owned by us all.

In a time when nature is threatened on every front, art reminds us of our humanity, of our relationship and responsibility to the world. Science has trained me to see beauty in discovery. Nature has shown me that art is her sister.

Me decís qué espera la ascidia en su campana transparente?
qué espera?
Yo os digo,
espera como vosotros el tiempo.

You ask what the luminous bell of the tunicate awaits in the water:
what does it hope for?
I tell you,
it waits for the fullness of time,
like yourself.

—Pablo Neruda, from *Los Enigmas*

INDEX